Characterizations of C(X) among Its Subalgebras

Lecture Notes in Pure and Applied Mathematics

1. *N. Jacobson*, Exceptional Lie Algebras
2. *L.-Å. Lindahl* and *F. Poulsen*, Thin Sets in Harmonic Analysis
3. *I. Satake*, Classification Theory of Semi-Simple Algebraic Groups
4. *F. Hirzebruch, W. D. Neumann*, and *S. S. Koh*, Differentiable Manifolds and Quadratic Forms
5. *I. Chavel*, Riemannian Symmetric Spaces of Rank One
6. *R. B. Burckel*, Characterizations of C(X) among Its Subalgebras

Other volumes in preparation

Characterizations of C(X) among Its Subalgebras

R. B. BURCKEL
Kansas State University
Manhattan, Kansas

MARCEL DEKKER, INC. New York 1972

MARCEL DEKKER, INC.

95 Madison Avenue, New York, New York 10016

LIBRARY OF CONGRESS CATALOG CARD NUMBER 72-90373

ISBN 0-8247-6038-7

PRINTED IN THE UNITED STATES OF AMERICA

To the Memory of my Father

PREFACE

These notes are a revised version of lectures I gave at the University of Oregon in the Spring of 1970. Since that time I have had access to several more recent papers, lecture notes and monographs and have freely used them to expand and improve my original account. Still, some of the material here has not appeared in monographic form before.

My aim is to present a detailed account of some recent results--almost all the material is less than a decade old--about subalgebras of $C(X)$. These algebras carry a Banach algebra norm which is often, but not always, the uniform norm. Accordingly, students of Banach algebra theory will perceive that we are really concerned with commutative semi-simple Banach algebras, but I do not use the language of or any non-elementary results about Banach algebras. In fact, nothing more recondite than the Gelfand spectral radius formula is used, and this is used only once, in one proof of a theorem for which two proofs are provided. I have made these self-imposed limitations because this is an evangelistic effort; I wish to communicate these beautiful results to as large a public as possible, especially students. The prospective reader should have had a standard graduate real-variable course and be acquainted with a few odds and ends of functional analysis and complex-variables. For example, the first 225 pages of Rudin's book [41] should be more than adequate equipment. Modulo this background, all proofs are complete and, in fact, some will find the proof detail excessive. However, my attitude here is that it is easier for the reader to skip over details of a proof which he understands than to supply these details if he does not.

I thank Irving Glicksberg for making the notes [22] available to me, Stuart Sidney for the notes [46] and some helpful correspondence and errata, and Lee Stout, William Badé, and Yitzhak Katznelson for helpful correspondence over their work. Especial thanks to Kenneth A. Ross for his meticulous reading of my account, to Carolyn Hobbs for typing it and to the original seminar members for critically listening to it.

R. B. Burckel

CONTENTS

Characterizations of C(X) among Its Subalgebras

SOME NOTATIONS AND GENERAL REMARKS

(I) We use $\mathbb{C}$ for the complexes, $\mathbb{R}$ for the reals, Z for the integers, $\mathbb{R}^+$ for the non-negative reals, Z^+ for the non-negative integers, $\mathbb{N}$ for the positive integers (viz. the natural numbers).

(II) If X is a topological space, we write $C(X)$ for the set of all bounded continuous complex-valued functions on X, and $C_0(X)$ for the subset of these functions which vanish at infinity in the sense of being arbitrarily small off compact sets: $f \in C_0(X)$ if and only if $f \in C(X)$ and $|f|^{-1}[\epsilon, \infty)$ is compact for each $\epsilon > 0$. We write $C_{\mathbb{R}}(X)$ for the set of real-valued functions in $C(X)$. If $Y \subset X$ and $f \in C(X)$, we write $f|Y$ for the restriction of f to Y and $E|Y$ for the set of restrictions to Y of the functions in E whenever $E \subset C(X)$. We write $\|f\|_\infty$ or $\|f\|_X$ for $\sup|f(X)|$. We also write $\|f\|_Y$ for $\|f|Y\|_\infty$. Any otherwise unqualified topological statements about $C(X)$ will refer to the norm topology generated by $\| \ \|_\infty$.

(III) If X is a topological space and A a Banach algebra lying in $C(X)$, then $\|f\|_\infty \le \|f\|_A$ for every $f \in A$. Indeed if $\lambda \in \mathbb{C}$ and $|\lambda| > \|f\|_A$, then $\|\lambda^{-1}f\|_A = r < 1$ and so for any positive integers m and p,

$$\left\|\sum_{n=m}^{m+p}(\lambda^{-1}f)^n\right\|_A \le \sum_{n=m}^{m+p}\|\lambda^{-1}f\|_A^n \le \sum_{n=m}^{m+p} r^n < \sum_{n=m}^{\infty} r^n = \frac{r^m}{1-r} .$$

Since A is complete, the series $\sum_{n=1}^{\infty}(\lambda^{-1}f)^n$ therefore converges in A, say to g. A trivial calculation shows that the function g satisfies $g - \lambda^{-1}f - \lambda^{-1}fg = 0$. It follows that $f(x) \neq \lambda$ for every $x \in X$. Thus $f(X) \subset \{\lambda \in \mathbb{C}: |\lambda| \le \|f\|_A\}$, that is, $\|f\|_\infty \le \|f\|_A$.

(IV) Our attitude toward restriction algebras and vector spaces is ambivalent: Let X be a topological space, $Y \subset X$, and $E \subset C(X)$ a normed linear space continuously injected in $C(X)$, that is, $\| \ \|_\infty \le M\| \ \|_E$ for some constant M. The set $kY = \{f \in E: f(Y) = 0\}$ is closed in the norm of E, because of this inequality. [The use of "k" here is to suggest kernel, as in the hull-kernel topology in Banach algebra theory.] The coset space E/kY and the restriction set $E|Y$ are clearly algebraically isomorphic vector spaces. This set will always be topologized by the quotient norm in E/kY and we shall treat it in whichever manifestation is more convenient for the problem at hand. Thus

$$\|f + kY\|_{E/kY} = \inf\{\|f + g\|_E: g \in kY\} \quad \text{for} \quad f \in E$$

and

$$\|h\|_{E|Y} = \inf\{\|f\|_E: f \in E,\ f|Y = h\} \quad \text{for} \quad h \in E|Y.$$

(a) When E is an algebra and $\| \ \|_E$ is submultiplicative, then this quotient is an algebra and this quotient norm is also submultiplicative, as is easy to see.

(b) If $\| \|_E$ is a complete norm, so is this quotient norm. The standard proof of this is as follows. Given $\{h_n\}$ Cauchy in $E|Y$ it suffices to show that some subsequence of $\{h_n\}$ has a limit. Passing to an appropriate subsequence, we may therefore suppose without loss of generality that $\|h_{n+1} - h_n\|_{E|Y} < 2^{-n}$. Thus there exist $f_n \in E$ with $f_n|Y = h_{n+1} - h_n$ and $\|f_n\|_E < 2^{-n}$. The series $\sum_{k=1}^{\infty} f_k$ therefore converges in E, say to f. We then have

$$\|f|Y + h_1 - h_n\|_{E|Y} = \|f|Y - \sum_{k=1}^{n-1} (h_{k+1} - h_k)\|_{E|Y} = \|(f - \sum_{k=1}^{n-1} f_k)|Y\|_{E|Y}$$

$$\le \|f - \sum_{k=1}^{n-1} f_k\|_E = \|\sum_{k=n}^{\infty} f_k\|_E \le \sum_{k=n}^{\infty} \|f_k\|_E$$

$$< \sum_{k=n}^{\infty} 2^{-k} = 2^{-n+1}.$$

It follows that $\{h_n\}$ converges in $E|Y$ to $h_1 + f|Y$.

(c) If $\| \|_\infty \le M\| \|_E$ then

$$\|h\|_\infty \le M\|h\|_{E|Y} \quad \text{for all } h \in E|Y.$$

Indeed, for any $f \in E$ with $f|Y = h$ we have

$$\|h\|_\infty \le \|f\|_\infty \le M\|f\|_E$$

and the infimum on the right over such f is exactly $M\|h\|_{E|Y}$.

(V) If X is a topological space and $E \subset C_0(X)$, we say E separates the points of X or is a point-separating set if for each $x \in X$ and $y \in X\backslash\{x\}$ there is an $f \in E$ such that $f(x) \neq f(y) \neq 0$.

(VI) All other notation is either standard, self-explanatory, or temporary and defined where introduced.

Chapter I

BISHOP'S STONE-WEIERSTRASS THEOREM

Definition 1.1

Let X be a compact Hausdorff space, A a (real or complex) subalgebra of (real or complex) $C(X)$ containing the constants. Call $K \subset X$ a set of antisymmetry of A or an A-antisymmetric set if $f \in A$ and f real on K implies f constant on K (i.e., $g, \bar{g} \in A|K$ implies g constant). Call A an antisymmetric algebra if X is an A-antisymmetric set, that is, if A contains no non-constant real functions.

With X and A as above note the following elementary facts:

1. If K_1, K_2 are A-antisymmetric sets and $K_1 \cap K_2 \neq \emptyset$ then $K_1 \cup K_2$ is A-antisymmetric. For if $x \in K_1 \cap K_2$, $f \in A$ and $f|K_1 \cup K_2$ is real-valued then $f|K_j$ are each constant so

$$f(K_1) = f(x) = f(K_2)$$

whence $f|K_1 \cup K_2$ is constant.

2. If $\emptyset \neq Y \subset X$ and $\mathcal{C}$ is a set of A-antisymmetric subsets of X which each contain Y then $\bigcup\mathcal{C}$ is again an A-antisymmetric set. For if $f \in A$ and f is real on $\bigcup\mathcal{C}$ then for each $K \in \mathcal{C}$, f is real on K whence constant on K and this constant must be $f(K) = f(Y)$, i.e. is the same for all $K \in \mathcal{C}$, so f is constant on $\bigcup\mathcal{C}$. Of course 1. above is a special case of this. Evidently then the union

of all the A-antisymmetric sets which contain Y is the maximal A-antisymmetric set which contains Y.

3. As each singleton $\{x\}$ is trivially an A-antisymmetric set, each x lies in a maximal A-antisymmetric set K.

4. Let K be a maximal A-antisymmetric set. If $f \in A$ is real on $\overline{K}$ then f is constant on K, whence on $\overline{K}$ by continuity. So maximality forces $K = \overline{K}$.

5. Finally by 1., distinct maximal A-antisymmetric sets must be disjoint.

This establishes the trivial parts of

Theorem 1.2

(Bishop [9]) Let X be a compact Hausdorff space, A a uniformly closed subalgebra of $C(X)$ which contains the constants. Let I be a uniformly closed ideal of A (which may be all of A). Then every A-antisymmetric set is contained in a maximal one. The collection $\mathcal{K}$ of maximal A-antisymmetric sets forms a pairwise disjoint closed cover of X satisfying:

(i) $f \in C(X)$ & $f|K \in I|K$ for each $K \in \mathcal{K} \Rightarrow f \in I$.

(ii) $I|K$ is uniformly closed in $C(K)$ for each $K \in \mathcal{K}$.

For (i) we need a lemma and some notation. $C(X)^* = M(X)$ is the finite regular Borel measures μ on X with the norm of μ as a functional on $C(X)$ equal to its total variation. $I^{\perp}$ is the annihilator of I in $C(X)^*$ and $(\text{ball } I^{\perp})^e$ is the extreme points of the unit ball of $I^{\perp}$. We denote the closed support of $\mu \in M(X)$ by $\operatorname{supp} \mu$, and for $f \in C(X)$

we write $\mu(f)$ for $\int_X f d\mu$ and $f\mu$ for the measure defined (as a linear functional) by $(f\mu)(h) = \mu(fh)$ for all $h \in C(X)$.

Lemma 1.3

(de Branges [11]) If $\mu \in (\text{ball } I^{\perp})^e$, then $\text{supp } \mu$ is a set of antisymmetry of A.

Proof: Let $K = \text{supp } \mu$, $f \in A$, $f(K) \subset \mathbb{R}$. To see f is constant on K it suffices (as A is a vector space containing 1) to suppose

(0) $\quad 0 < f < 1$ on K.

Then $f\mu$ and $(1 - f)\mu$ are non-zero measures so

(1) $$\mu = \|f\mu\| \frac{f\mu}{\|f\mu\|} + \|(1 - f)\mu\| \frac{(1 - f)\mu}{\|(1 - f)\mu\|}$$

while

(2) $$\begin{aligned}\|f\mu\| + \|(1 - f)\mu\| &= \int_X |f| d|\mu| + \int_X |1 - f| d|\mu| \\ &= \int_K |f| d|\mu| + \int_K |1 - f| d|\mu| \\ &= \int_K f d|\mu| + \int_K (1 - f) d|\mu| \qquad \text{by (0)} \\ &= \int_K 1 d|\mu| = |\mu|(K) = |\mu|(X) = \|\mu\| \\ &= 1.\end{aligned}$$

[For the first equality the reader may wish to consult Rudin [41], p. 126, and for the last equality recall that μ is extreme in the unit ball of $I^{\perp}$ so μ is not in the interior of that ball, i.e. $\|\mu\| \not< 1$ so $\|\mu\| = 1$.] Now $f\mu$ and $(1 - f)\mu$ lie in $I^{\perp}$ since for $g \in I$

$$(f\mu)(g) = \mu(fg) = 0 \qquad (fg \in I \text{ ideal}, \ \mu \in I^{\perp})$$

$$[(1-f)\mu](g) = \mu((1-f)g) = 0 \qquad ((1-f)g \in I \text{ ideal}, \ \mu \in I^{\perp}).$$

So by (1), (2) and the extremality of μ we get

$$\mu = \frac{f\mu}{\|f\mu\|} = \frac{(1-f)\mu}{\|(1-f)\mu\|}.$$

In particular, then, we have $|\mu|$ - a.e.

$$\frac{f}{\|f\mu\|} = \frac{1-f}{\|(1-f)\mu\|}$$

whence, setting $c_1 = \|f\mu\|/\|(1-f)\mu\|$ and $c_2 = 1 + c_1$,

(3) $\qquad f = c_1/c_2 \qquad\qquad |\mu|$ - a.e.

As f is continuous, the set $V = \{x \in X : f(x) \neq c_1/c_2\}$ is open. By (3), V has $|\mu|$-measure 0 and therefore by definition of the support of a measure, V is disjoint from supp $\mu = K$, i.e. f has the constant value c_1/c_2 throughout K.

Proof of (i) of Theorem 1.2: If $f \in C(X)$ and $f \notin I$, then by Hahn-Banach f is not annihilated by $I^{\perp}$, equivalently f is not annihilated by ball $I^{\perp}$. Then by Krein-Milman f is not annihilated by $(\text{ball } I^{\perp})^e$. For $\{f\}^{\perp} \cap \text{ball } I^{\perp}$ is a weak * compact convex set and if it contains $(\text{ball } I^{\perp})^e$, then it contains the weak * closed convex hull of $(\text{ball } I^{\perp})^e$, which by the Krein-Milman Theorem is all of ball $I^{\perp}$. Let $\mu \in (\text{ball } I^{\perp})^e$ be such that $\mu(f) \neq 0$. By the lemma, supp μ lies in some $K \in \mathcal{K}$. So for some $g \in I$, $f = g$ on K (hypothesis on f) and then

$$\mu(f) = \int_K f\,d\mu = \int_K g\,d\mu = \mu(g) = 0, \text{ as } \mu \in I^{\perp}.$$

This however contradicts the choice of μ.

Corollary 1.4

(Stone-Weierstrass) If X is a compact Hausdorff space and A is a uniformly closed subalgebra of $C(X)$ which separates the points of X, contains the constants and is conjugate closed, then $A = C(X)$.

Proof: Evidently, since A separates points and is conjugate closed, $A \cap C_{\mathbb{R}}(X)$ separates points. But then each $K \in \mathcal{K}$ is a singleton, so $f|K \in A|K$ for every $f \in C(X)$ and the theorem (with $I = A$) yields $C(X) \subset A$.

Note: Lest the reader sense circularity in this deduction of the traditional Stone-Weierstrass Theorem from Bishop's Theorem, let us point out that the Riesz Representation Theorem, the Hahn-Banach Theorem and the Krein-Milman Theorem, upon which alone Bishop's Theorem depends, are proven without recourse to the Stone-Weierstrass Theorem. Because of the way Stone-Weierstrass permeates analysis, such a fear would not prima facie have been ill-founded.

For (ii) of the Bishop Theorem we need some more lemmas.

Call $E \subset X$ a peak set of A or an A-peak set if $E = \{x \in X : f(x) = 1\}$ for some $f \in A$ with $\|f\|_\infty = 1$. Notice that if we set $F = \frac{1}{2}(1 + f)$ then $F \in A$, $\|F\|_\infty = 1$ and $|F(x)| = 1$ iff $F(x) = f(x) = 1$, so $E = \{x \in X : F(x) = 1\}$ and $|F(x)| < 1$ for each $x \notin E$. Let us say that an F with this last property peaks on E.

Note: A countable intersection of A-peak sets E_n is

also an A-peak set. For if $F_n \in A$ peaks on E_n, then clearly $\sum_{n=1}^{\infty} 2^{-n} F_n$ peaks on $\bigcap_{n=1}^{\infty} E_n$ and belongs to A.

Lemma 1.5

If E is an intersection of A-peak sets, then the uniform norm in $I|E$ equals the quotient norm. Since the quotient is a Banach space, it follows that $I|E$ is complete, hence closed in $C(E)$.

Proof: Recall (IV (c), p. 3) that for all $f \in A$

$$\|f|E\|_\infty \le \|f + I \cap kE\|.$$

On the other hand, if V_ϵ is the open set $\{x \in X: |f(x)| < \|f|E\|_\infty + \epsilon\}$, then some finite intersection E_ϵ of the A-peak sets containing E lies entirely in V_ϵ by compactness. As noted above, E_ϵ is again an A-peak set, so there exists $g_\epsilon \in A$ with $g_\epsilon(E_\epsilon) = 1$ and $|g_\epsilon(x)| < 1$ for all $x \notin E_\epsilon$. On the compact set $X\backslash V_\epsilon$ then g_ϵ has a supremum less than 1, so for large n

$$\sup|g_\epsilon^n f(X\backslash V_\epsilon)| < \sup|f(V_\epsilon)| + \epsilon.$$

It follows that

$$\sup|g_\epsilon^n f(X)| \le \max\{\sup|f(V_\epsilon)|+\epsilon, \sup|g_\epsilon^n f(V_\epsilon)|\} = \sup|f(V_\epsilon)|+\epsilon.$$

Consulting the definition of V_ϵ we see $\sup|f(V_\epsilon)| \le \|f|E\|_\infty + \epsilon$, so we've proved

$$\sup|g_\epsilon^n f(X)| = \|g_\epsilon^n f\|_\infty \le \|f|E\|_\infty + 2\epsilon.$$

Since $g_\epsilon^n f - f = (g_\epsilon^n - 1)f = 0$ on $E_\epsilon \supset E$, we see that $g_\epsilon^n f - f \in kE$. Also, I is an ideal, so $f \in I$ implies

$g_\epsilon^n f - f \in I$. Therefore $g_\epsilon^n f - f \in I \cap kE$ and our last inequality yields

$$\|f + I \cap kE\| \le \|g_\epsilon^n f\|_\infty \le \|f|E\|_\infty + 2\epsilon.$$

As ϵ is arbitrary, we're done.

Lemma 1.6

(Bishop [10]) If $L \subset F \subset X$, F is an intersection of A-peak sets and L is an $A|F$-peak set, then $L = S \cap F$ for some A-peak set S. If in particular F is an A-peak set, then so is L.

Proof: Let $f \in A$ be such that $f|F$ peaks on L, that is,

$$L = \{x \in F: f(x) = 1\} \text{ and } |f(x)| < 1 \text{ for each } x \in F \backslash L.$$

For each positive integer n set

$$G_n = \{x \in X: |f(x)| < 1 + \frac{1}{2^n}\}.$$

This is an open set containing F and so some finite number of the A-peak sets containing F have their intersection in G_n, by compactness. This finite intersection is again an A-peak set, so for some $h_n \in A$ we have

(1) $$F \subset \{x \in X: h_n(x) = \|h_n\|_\infty = 1\} = \{x \in X: |h_n(x)| = 1\} \subset G_n.$$

As the continuous function $|h_n|$ is then less than 1 on the compact set $X \backslash G_n$, we have

$$\sup|h_n(X \backslash G_n)| = \delta_n < 1.$$

Choose positive integer k_n so that

$$\delta_n^{k_n} < [2^n(\|f\|_\infty + 1)]^{-1}.$$

Then

$$(2)\quad |h_n^{k_n}(x)f(x)| \le \delta_n^{k_n}|f(x)| < \delta_n^{k_n}(\|f\|_\infty + 1) < 2^{-n} \qquad \forall x \in X\backslash G_n.$$

Consider $g = f\sum_{n=1}^{\infty} 2^{-n}h_n^{k_n}$. Note that since $\|h_n\|_\infty = 1$, the series converges uniformly (so $g \in A$) and

$$|g(x)| \le |f(x)| \sum_{n=1}^{\infty} 2^{-n}|h_n(x)|^{k_n} \le |f(x)| \qquad \forall x \in X.$$

Therefore if $x \in \bigcap_{n=1}^{\infty} G_n$ then $|f(x)| \le 1$ and so

$$(3)\quad |g(x)| \le 1 \qquad \forall x \in \bigcap_{n=1}^{\infty} G_n.$$

If $x \notin \bigcap_{n=1}^{\infty} G_n$ then, defining $G_0 = X$, there exists $m \ge 1$ for which $x \in G_{m-1}$ while $x \notin G_n$ for all $n \ge m$. Thus

$$|h_n^{k_n}(x)f(x)| < 1 + \frac{1}{2^{m-1}} \quad \text{for} \quad n = 1,2,\ldots,m-1$$

since $\|h_n\|_\infty = 1$, while $|f| < 1 + \frac{1}{2^{m-1}}$ on G_{m-1}. Also

$$|h_n^{k_n}(x)f(x)| < \frac{1}{2^n} < \frac{1}{2^{m-1}} \quad \text{for} \quad n \ge m$$

by (2) and the fact $x \notin G_n$. Therefore

$$(4)\quad |g(x)| \le \left(1 + \frac{1}{2^{m-1}}\right)\sum_{n=1}^{m-1}\frac{1}{2^n} + \frac{1}{2^{m-1}}\sum_{n=m}^{\infty}\frac{1}{2^n}$$

$$= \left(1 + \frac{1}{2^{m-1}}\right)\left(1 - \frac{1}{2^{m-1}}\right) + \frac{1}{2^{m-1}} \cdot \frac{1}{2^{m-1}} = 1 \qquad \forall x \notin \bigcap_{n=1}^{\infty} G_n.$$

From (3) and (4) we get

$$\|g\|_\infty \le 1.$$

Therefore, if $S = g^{-1}(1)$, we have that S is a peak set of A. Since each h_n is 1 on F by(1), we have $g = f$ on F and so

$$S \cap F = \{x \in F\colon g(x) = 1\} = \{x \in F\colon f(x) = 1\} = L.$$

Lemma 1.7

Every maximal A-antisymmetric set K is an intersection of A-peak sets.

Proof: There are peak sets which contain K, for example X. Let E denote the intersection of all of them. We will show that $K = E$. If not, then by maximality of K, E is not a set of A-antisymmetry so there exists an f in A with $f|E$ real but not constant. $f(E)$ being a compact subset of $\mathbb{C}$, there exist U_n open subsets of $\mathbb{C}$ such that $f(E) = \bigcap_{n=1}^{\infty} U_n$. (E.g., let $U_n = \{z \in \mathbb{C}: d(z,f(E)) < \frac{1}{n}\}$.) Then $f^{-1}(f(E)) = \bigcap_{n=1}^{\infty} V_n$, where V_n is the open set $f^{-1}(U_n)$. By compactness, $E \subset V_n$ implies that some finite intersection E_n of the peak sets determining E already lies in V_n. Thus $E \subset F \overset{\text{def}}{=} \bigcap_{n=1}^{\infty} E_n \subset \bigcap_{n=1}^{\infty} V_n \subset f^{-1}(f(E))$, and so $f(E) = f(F)$. Moreover, as noted above, F is again a peak set.

Since $f(E) = f(F)$ and f is real and not constant on E, f is real and not constant on F. But f being real in $E \supset K$, is constant on the antisymmetry set K, say $f(K) = c$. We therefore have

$$a = \min f(F) < b = \max f(F).$$

$$c \in [a,b].$$

Pick N large enough that $[c - N, c + N] \supset [a,b]$ and consider the polynomial

$$p(t) = 1 - [\frac{t-c}{N}]^2.$$

One checks that

$$0 \le p(t) \le p(c) = 1 \qquad \forall t \in [c - N,\ c + N]$$

and that c is the unique maximizing point in this interval.

Consider now $p \circ f \in A$. We have that

$$f(F) \subset [a,b] \subset [c - N,\ c + N]$$

and so

$$(p \circ f)(K) = p(f(K)) = p(c) = 1 = \max p[c-N,c+N] = \max p[a,b] \ge \max p(f(F)).$$

Therefore if L denotes the subset of F where $p \circ f$ is 1 we see that L is a peak set for $p \circ f|F$ and that $K \subset L$. Moreover $a < b$ and both $a,b \in f(F)$ so either $a \neq c$ or $b \neq c$, whence either $p(a) < p(c)$ or $p(b) < p(c)$. So at least one point of F is not in L. As we have now shown that L is a peak set for $A|F$, the last lemma insures that L is a peak set for A. But then the fact $L \supset K$ and the definition of E imply $E \subset L$. So $p \circ f(E) \subset p \circ f(L) = \{1\}$ and since $f(E) = f(F)$ this gives

$$p(f(F)) = p(f(E)) \subset \{1\}.$$

This contradicts the fact that the peak set $L = (p \circ f)^{-1}(1) \cap F$ is a proper subset of F.

<u>Proof of (ii) of Theorem 1.2</u>: It remains only to combine lemmas 1.5 and 1.7 to conclude that $I|K$ is uniformly closed in $C(K)$.

<u>Corollary 1.8</u>

Let X be a compact Hausdorff space, A a uniformly closed subalgebra of $C(X)$ which contains the constants, I a uniformly closed ideal of A. If $f \in I$ and $\overline{f} \in A$, then

$\overline{f} \in I$. That is, uniformly closed ideals are conjugate closed to the extent that A itself is.

<u>Proof</u>: Let $\mathcal{K}$ be the maximal A-antisymmetric decomposition of X. $f + \overline{f}$ and $i(\overline{f} - f)$ are real-valued functions in A and so they are constant on each $K \in \mathcal{K}$. Then $f = \frac{1}{2}(f + \overline{f}) - \frac{1}{2}(\overline{f} - f)$ is constant on K, say $f = c_K$ on K. If $c_K = 0$ then $\overline{f}|K = c_K = 0 \in I|K$, while if $c_K \neq 0$ $\overline{f}|K = \overline{c}_K = (\frac{1}{c_K}\overline{f})\cdot f|K \in I|K$, since $f \in I$, $\overline{f} \in A$ and I is an ideal. Now quote Theorem 1.2.

<u>Remark</u>. This proof of Bishop's Theorem and its corollaries is from Glicksberg [20].

Chapter II

RESTRICTION ALGEBRAS DETERMINING C(X)

Our next result seems to be due independently to Chalice [14], Mullins [38] and Wilken & Gamelin [52]. Compare also [30]. The proof below is adapted from [38].

Theorem 2.1

Let X be a compact Hausdorff space, A a uniformly closed, point-separating subalgebra of C(X) which contains the constants. Suppose $F_1, F_2, \ldots$ is a sequence of closed subsets of X such that $\bigcup_{j=1}^{\infty} F_j = X$ and $A|F_j = C(F_j)$ for each j. Then A = C(X).

It is convenient to introduce a term and prove some lemmas first.

Definition 2.2

Let X be a compact Hausdorff space, A a subalgebra of C(X). Call $x \in X$ a strong A-boundary point if each neighborhood of x contains an A-peak set which contains x. Let ∂A denote the closure of the set of strong A-boundary points.

Lemma 2.3

(Runge) Let $0 < r_0 < r_1 < 1$ and $B = \{z \in \mathbb{C}: |z| \le r_0\}$, $K = B \cup \{z \in \mathbb{C}: |1-z| \le 1-r_1\}$. Then on K the characteristic function χ_B of B is a uniform limit of polynomials.

<u>Proof</u>: Clearly $\chi_B \in C(K)$, since B is clopen in K. By Hahn-Banach we have only to show that if $\mu \in C(K)^* = M(K)$ annihilates all polynomials then μ annihilates χ_B. So we suppose

$$(1) \qquad \int_K \xi^n d\mu(\xi) = 0 \qquad n = 0,1,2,\ldots.$$

Define the function $\hat{\mu}$ on $\mathbb{C}\backslash K$ by

$$(2) \qquad \hat{\mu}(z) = \int_K \frac{1}{z-\xi}\, d\mu(\xi) \qquad \forall z \notin K.$$

If, say, $|z| \geq 2$ then the series

$$\frac{1}{z} \sum_{n=0}^{\infty} \left(\frac{\xi}{z}\right)^n$$

converges uniformly for $\xi \in K$ to $\frac{1}{z} \cdot \frac{1}{1-\frac{\xi}{z}} = \frac{1}{z-\xi}$

(for $\xi \in K$ implies $|\xi| \leq 2-r_1$ implies $|\frac{\xi}{z}| \leq \frac{2-r_1}{2} < 1$).

Therefore for such z

$$\hat{\mu}(z) = \int_K \frac{1}{z} \sum_{n=0}^{\infty} \left(\frac{\xi}{z}\right)^n d\mu(\xi) = \sum_{n=0}^{\infty} \frac{1}{z} \int_K \left(\frac{\xi}{z}\right)^n d\mu(\xi)$$

$$= \sum_{n=0}^{\infty} \frac{1}{z^{n+1}} \int_K \xi^n d\mu(\xi) = 0 \qquad \text{by (1).}$$

Thus we have

$$(3) \qquad \hat{\mu}(z) = 0 \qquad \forall |z| \geq 2.$$

Now $\hat{\mu}$ is analytic in the open set $\mathbb{C}\backslash K$, for it is easy to justify differentiating under the integral (2) with respect to z (in $\mathbb{C}\backslash K$). Moreover the set $\mathbb{C}\backslash K$ is obviously connected. Therefore from (3) and the uniqueness Theorem of Analytic Function Theory we conclude

$$(4) \qquad \hat{\mu}(z) = 0 \qquad \forall z \in \mathbb{C}\backslash K.$$

Let $r_0 < r < r_1$. Then

$$\frac{1}{2\pi i}\int_0^{2\pi}\frac{rie^{it}}{re^{it}-z}\,dt = \begin{cases} 1 \text{ if } |z| < r & \text{(Cauchy's Integral Formula for a disk)} \\ 0 \text{ if } |z| > r & \text{(Cauchy's Theorem for a disk).} \end{cases}$$

In particular

$$(5) \qquad \chi_B(z) = \frac{1}{2\pi}\int_0^{2\pi}\frac{re^{it}}{re^{it}-z}\,dt \qquad \forall z \in K.$$

Now the function $(\xi,z) \to \frac{\xi}{\xi-z}$ is continuous on $\{re^{it}: 0 \le t \le 2\pi\} \times K$ and so the simplest form of Fubini gives

$$2\pi\int_K \chi_B(z)d\mu(z) \overset{(5)}{=} \int_K\int_{[0,2\pi]}\frac{re^{it}}{re^{it}-z}\,dtd\mu(z)$$

$$= \int_{[0,2\pi]}\int_K \frac{re^{it}}{re^{it}-z}\,d\mu(z)dt$$

$$= \int_{[0,2\pi]}re^{it}\int_K \frac{1}{re^{it}-z}\,d\mu(z)dt.$$

But $\{re^{it}: 0 \le t \le 2\pi\} \subset \mathbb{C}\backslash K$, so by (4) and (2) the inner integral above is identically zero and we're done.

Lemma 2.4

Let X be a compact Hausdorff space, A a uniformly closed point-separating subalgebra of $C(X)$ which contains the constants. Then every A-peak set contains a strong A-boundary point.

Proof: Zornicate. In more detail, let E be an A-peak set, say $E = \{x \in X: |f(x)| = \|f\|_\infty\}, f \in A$. Evidently if $\mathcal{C}$ is a linearly ordered (by inclusion) chain of non-void intersections of A-peak sets lying in E then $\cap\mathcal{C}$ is non-void (by compactness of E) and an intersection of A-peak sets. Therefore Zorn's lemma implies that E contains a non-void set F which is an intersection of A-peak sets and does not properly contain any such non-void intersection. We contend that F is a single point. For if F contains more than one point, A will contain a function g which is not constant on F, by the point-separation hypothesis. In particular $g|F \neq 0$. Pick $x_0 \in F$ such that $0 < |g(x_0)| = \|g|F\|_\infty$ and consider $h = \frac{1}{2}(1 + \frac{g}{g(x_0)}) \in A$. The set $L = \{x \in F: h(x) = 1\}$ is a proper (since g is not constant on F), non-void (since x_0 belongs) subset of F which is an $A|F$-peak set. By lemma 1.6, L is an intersection of A-peak sets and the minimality of F is compromised. We conclude that $F = \{x\}$ for some x. If V is any open neighborhood of x then some finite number of the A-peak sets containing F have their intersection in V, by compactness. This finite intersection is again an A-peak set, and so x is a strong A-boundary point.

Lemma 2.5

With the notation and hypotheses of the last lemma, suppose A is an antisymmetric algebra and that some strong A-boundary point has an open neighborhood V such that $A|\overline{V} = C(\overline{V})$. Then X is a single point.

Proof: Let x_0 be a strong A-boundary point in V and let $g \in A$ be such that $\{x \in X: g(x) = g(x_0) = \|g\|_\infty = 1\} \subset V$ and $|g(x)| < 1$ for $x \notin V$. So there exists $0 < r_0 < 1$ such that

$$|g(X\backslash V)| < r_0.$$

Pick $r_0 < r_1 < 1$ and set

$$B = \{z \in \mathbb{C}: |z| < r_0\}, \quad D = \{z \in \mathbb{C}: |z-1| < 1-r_1\}.$$

By lemma 2.3 there are polynomials p_n such that $p_n \to \chi_B$ uniformly on $B \cup D$. If we set

$$U = \{x \in X: |g(x) - 1| < 1-r_1\},$$

then U is an open subset of X which contains x_0 and lies in V. Let f be any element of $C(X)$ which vanishes outside U. Since $A|\overline{V} = C(\overline{V})$, we can pick $h \in A$ with $h|\overline{V} = f|\overline{V}$. Now we have

$$g(U) \subset D \quad \text{and} \quad g(X\setminus V) \subset B,$$

so the sequence $\{p_n \circ g\}$ converges uniformly on $U \cup (X\setminus V)$ to the characteristic function of U. Therefore, since h vanishes on $V\setminus U$, $\{h\cdot(p_n\circ g)\}$ converges uniformly on X to $h\cdot\chi_U$. But $h\cdot\chi_U = f$ since f vanishes outside U and $h = f$ on $V \supset U$. Since $h\cdot(p_n\circ g) \in A$, the uniform limit f of this sequence is in A.

If U were not all of X there would be a non-constant real-valued $f \in C(X)$ vanishing off U. This f would belong to A, as just shown, and the antisymmetry of A would be violated. We conclude that $U = X$. Since $U \subset V$, we have finally $\overline{V} = X$, whence $C(X) = C(\overline{V}) = A|\overline{V} = A$ and because of the antisymmetry and point-separation properties of A we see that X must reduce to a single point.

Lemma 2.6

Under the hypotheses of Theorem 2.1 and the additional hypothesis that A is antisymmetric and $\partial A = X$ it follows that X is a single point.

Proof: By the Baire Category Theorem some F_j has non-void interior V_j. The hypothesis $\partial A = X$ means that the strong A-boundary points are dense in X so there is one in

V_j. The asserted conclusion follows then from the last lemma.

Lemma 2.7

Under the hypotheses of Theorem 2.1 and the additional hypothesis that A is antisymmetric it follows that X is a single point.

Proof: We first show that ∂A is an A-antisymmetric set. Suppose not. Then there exists $g \in A$ such that $g(\partial A)$ is real but g is not constant on ∂A. As X is by hypothesis an A-antisymmetric set, it must be that g is not real on all of X. Let, say, $\operatorname{Im} g(x_0) = \beta < 0$ for some $x_0 \in X$. The function $f = \exp\circ(ig)$ belongs to A and since g is real on ∂A it satisfies

$$|f(\partial A)| = 1$$
$$|f(x_0)| = e^{-\operatorname{Im} g(x_0)} = e^{-\beta} > 1.$$

Therefore $\|f\|_\infty > 1$ and $E = \{x \in X: |f(x)| = \|f\|_\infty\}$ contains an A-peak set disjoint from ∂A. By the definition of ∂A this means that E contains no strong A-boundary points, in contradiction to lemma 2.4.

Another application of lemma 2.4 shows that every function in A attains its maximum modulus on ∂A and so the map $f \to f|\partial A$ is an isometry. Since A is uniformly closed in $C(X)$, we see that $A|\partial A$ is uniformly closed in $C(\partial A)$. Of course $A|F_j = C(F_j)$ implies $(A|\partial A)|(F_j \cap \partial A) = C(F_j \cap \partial A)$ for each j. Finally it is clear that $\partial(A|\partial A) = \partial A$ and we are at last in a position to apply lemma 2.6 to the restriction algebra $A|\partial A$. It follows that ∂A is a single point. Then as $f \to f|\partial A$ is 1-1, it must be that A consists of constants.

Since A separates the points of X, this means that X is a single point.

Proof of Theorem 2.1: If $\mathcal{K}$ is Bishop's maximal A-antisymmetric decomposition of X, then for each $K \in \mathcal{K}$, $A|K$ is closed (Bishop's Theorem) and satisfies the hypotheses of lemma 2.7 with $K \cap F_j$ in the role of F_j. Conclude that K is a single point for each $K \in \mathcal{K}$, hence by Bishop's Theorem $A = C(X)$.

Corollary 2.8

If Y is a compact, countable subset of $\mathbb{C}$ then every continuous complex function on Y is a uniform limit of polynomials.

Proof: Write $Y = \{y_1, y_2, \ldots\}$, set $F_j = \{y_j\}$, let A be the uniform closure in $C(Y)$ of the polynomials (in the complex variable z) and apply Theorem 2.1 to conclude $A = C(Y)$.

Of course Corollary 2.8 is just the special case of Corollary 2.9 below needed to prove the latter:

Corollary 2.9

(Rudin [48]) Let X be a compact Hausdorff space without non-void perfect subsets, A a uniformly closed, point-separating subalgebra of $C(X)$ with $1 \in A$. Then $A = C(X)$.

Proof: Let $f \in A$, $Y = f(X)$. We will show that this compact set is countable. The last corollary then provides a sequence of polynomials p_n such that $\sup_{z \in Y} |p_n(z) - \bar{z}| \to 0$. It follows that $\|p_n \circ f - \bar{f}\|_\infty \to 0$ and so $\bar{f} \in A$. Thus A is conjugate closed and so the conclusion follows from Stone-Weierstrass

Let $V_1, V_2, \ldots$ be an enumeration of the open disks with rational radii and centers having rational coordinates which intersect Y in finite or countable sets. Then $C = Y \cap \bigcup_{n=1}^{\infty} V_n$ is countable and $P = Y \setminus C$ is closed and has no isolated points. For if $x \in P$ is isolated, there is a disc V of rational radius and center having rational coordinates such that $V \cap P = \{x\}$. But then $V \cap Y = (V \cap P) \cup (V \cap C) \subset \{x\} \cup C$ is countable and so by definition V is among the V_n, whence $x \in V \cap Y \subset C$, a contradiction. We are therefore finished if we show that P is void and we argue this by contradiction, showing that if $P \neq \emptyset$ then P has an isolated point. To this end we want a minimal f pre-image for P and, of course, Zorn provides one: if $\mathcal{C}$ is a linearly ordered (by inclusion) set of compact subsets K of X such that $f(K) = P$, then for each $y \in P$, $f^{-1}(y) \cap K \neq \emptyset$ for each $K \in \mathcal{C}$ so by compactness (and the linearity of the inclusion order on $\mathcal{C}$), $\emptyset \neq \bigcap_{K \in \mathcal{C}} f^{-1}(y) \cap K = f^{-1}(y) \cap \bigcap_{K \in \mathcal{C}} K$. That is, $f(\bigcap_{K \in \mathcal{C}} K) = P$. There is therefore in the family of all compact f pre-images of P a minimal one E. By hypothesis on X, the set E is not perfect and so has an isolated point x_0. Then $E \setminus \{x_0\}$ is closed and so by minimality of E we must have $f(E \setminus \{x_0\}) \neq P = f(E)$, that is, $P \setminus \{f(x_0)\} = f(E) \setminus \{f(x_0)\} = f(E \setminus \{x_0\})$. As $E \setminus \{x_0\}$ is compact, so is $f(E \setminus \{x_0\}) = P \setminus \{f(x_0)\}$, and so $f(x_0)$ is isolated in P, our final contradiction.

We proceed next to locally compact generalizations of Theorem 2.1. The first result in this direction, Theorem 2.10, is from the paper of Badé and Curtis in [8], pp. 90-92; they attribute the proof to Katznelson. A generalization of Theorem 2.1 of a different kind will appear in chapter six (Corollary 6.15).

Theorem 2.10

Let Y be a locally compact Hausdorff space, A a uniformly closed subalgebra of $C_0(Y)$ such that $A|F = C(F)$ for every compact $F \subset Y$. Then $A = C_0(Y)$.

Proof: It suffices to show that every continuous function f on Y which has compact support belongs to A. For A is uniformly closed and (as an easy consequence of Urysohn's lemma) every function in $C_0(Y)$ is a uniform limit of compactly supported functions.

Let K be a compact set off which f vanishes. Urysohn provides a $\varphi \in C_0(Y)$ with $\varphi(Y) \subset [0,1]$, $\varphi(K) = 1$ and φ supported in a compact $S \supset K$. Let V be an open neighborhood of S with compact closure and pick $g \in A$ such that $g|\overline{V} = \varphi|\overline{V}$. Let $F = |g|^{-1}[\frac{1}{2},\infty)\backslash V$. Since $g \in C_0(Y)$, the set $|g|^{-1}[\frac{1}{2},\infty)$ is compact, so F is compact. Now $S \cup F$ is compact and $S = V \cap (S \cup F)$ is open therein. Therefore the characteristic function χ_S of S belongs to $C(S \cup F)$. By hypothesis there is then an $h \in A$ such that $h|S \cup F = \chi_S f$. Now for each positive integer n, $g^n h \in A$ and we have

$$\|g^n h - f\|_{Y\backslash V \cup F} = \|g^n h\|_{Y\backslash V \cup F} \text{ since } f \text{ is supported in } K \subset V,$$

(1) $$\leq 2^{-n}\|h\|_Y \text{ since } V \cup F \supset |g|^{-1}[\tfrac{1}{2},\infty).$$

$$\|g^n h - f\|_F = \|g^n h\|_F \text{ since } f \text{ is supported in } K \subset V$$
which is disjoint from F (definition of F),

(2) $\qquad = 0$ since $h(F) = 0$.

$\|g^n h - f\|_{V \setminus S} = \|g^n h\|_{V \setminus S}$ since f is supported in $K \subset S$

(3) $\qquad = 0$ since $g = \varphi$ in V and φ is supported in S.

$\|g^n h - f\|_S = \|g^n f - f\|_S$ since $h = f$ in S,

(4) $\qquad = 0$ since $g = \varphi$ in $V \supset S$ and $\varphi = 1$ on the support of f.

It follows from (1)-(4) that $\|g^n h - f\|_Y \leq 2^{-n}\|h\|_Y \to 0$, so $f \in A$ since each $g^n h \in A$ and A is closed.

Lemma 2.11

Let X be a locally compact Hausdorff space, A a uniformly closed point-separating subalgebra of $C_0(X)$, $F_1, F_2, \ldots$ a sequence of compact subsets of X such that $\bigcup_{j=1}^{\infty} F_j = X$ and $A|F_j = C(F_j)$ for each j. Then $A = C_0(X)$.

Proof: Let $\tilde{X}$ be the one point compactification of X, x_0 the point adjoined, $F_0 = \{x_0\}$. Let A_0 be the functions in A extended to $\tilde{X}$ as 0 at x_0, and set $\tilde{A} = \mathbb{C} + A_0$. Then $\tilde{A}$ is evidently a uniformly closed, point-separating subalgebra of $C(\tilde{X})$ which contains the constants. Moreover

$$\tilde{A}|F_0 = \mathbb{C} + A_0|F_0 = C = C(F_0)$$
$$\tilde{A}|F_j = \mathbb{C} + A_0|F_j = C + A|F_j = C(F_j) \qquad j = 1,2,\ldots$$

and the F_j $(j = 0,1,2,\ldots)$ are compact subsets of $\tilde{X}$ which cover $\tilde{X}$. Therefore Theorem 2.1 ensures that $\tilde{A} = C(\tilde{X})$. It follows easily that A_0 comprises all the functions in $C(\tilde{X})$ which vanish at x_0 and so $A = C_0(X)$.

Corollary 2.12

Let Y be a locally compact Hausdorff space, A a uniformly closed, point-separating subalgebra of $C_0(Y)$. Let $\{Y_\alpha\}$ be a family of closed subsets of Y such that $A|Y_\alpha = C_0(Y_\alpha)$ for every α. If each point of Y has a neighborhood which is covered by countably many Y_α, then $A = C_0(Y)$.

Proof: By Theorem 2.10 it suffices to show that $A|F = C(F)$ for each compact $F \subset Y$. At every point of such an F some function in A is not zero. Then by compactness of F there exist $f_1,\ldots,f_n \in A$ such that F is covered by the open sets $V_k = |f_k|^{-1}(0,\infty)$, $k = 1,2,\ldots,n$. Set

(1) $\quad X = V_1 \cup \ldots \cup V_n.$

Each V_k is σ-compact, since $f_k \in C_0(Y)$. The hypothesis and a covering argument show that every σ-compact subset of Y is covered by countably many Y_α. It follows that we can write

(2) $\quad X = \bigcup_{m=1}^{\infty} F_m$

where each F_m is a compact set such that

(3) $\quad F_m \subset V_k \cap Y_\alpha$

for some k and some α. In particular, we then have

(4) $\quad A|F_m = (A|Y_\alpha)|F_m = C_0(Y_\alpha)|F_m = C(F_m).$

Given $\varphi \in C(F_m)$ we have, with k as in (3), $|f_k| > 0$ on $V_k \supset F_m$. Thus $\varphi/f_k|F_m \in C(F_m)$ and by (4) there exists $f \in A$ with $f|F_m = \varphi/f_k|F_m$. Therefore $f_kf|F_m = \varphi$. Now f_kf vanishes outside $V_k \subset X$. If therefore we set $I = \{f \in A\colon f(Y\backslash X) = 0\}$, a closed ideal in A, we shall have

(5) $\quad I|F_m = C(F_m) \qquad m = 1,2,\ldots$.

Evidently $I|X$ is a uniformly closed subalgebra of $C_0(X)$. We will show that I separates the points of X and then, in the light of (2) and (5), $I|X$ will satisfy the hypotheses of lemma 2.11. Indeed, if $x \in X$ then $x \in V_k$ for some k, so $|f_k(x)| > 0$ and $f_k \in I$. If also $y \in X$ and $y \neq x$, there are two possibilities. If $y \notin V_k$ then $f_k(y) = 0$, so $f_k(y) \neq f_k(x)$. If $y \in V_k$ then either $f_k(x) \neq f_k(y)$ or else $f_k(x) = f_k(y) \neq 0$. In the latter case, because A separates points, there is an $f \in A$ with $f(x) \neq f(y)$. Then $g = ff_k \in I$ and $g(x) \neq g(y)$.

Apply lemma 2.11 to learn that $I|X = C_0(X)$. Since F is a compact subset of X it follows that

$$A|F \supset I|F = (I|X)|F = C_0(X)|F = C(F).$$

<u>Corollary 2.13</u>

Let X be a compact Hausdorff space, A a uniformly closed, point-separating subalgebra of $C(X)$ which contains the constants. Suppose that each $x \in X$, with at most finitely many exceptions, has a compact neighborhood F_x such that $A|F_x = C(F_x)$. Then $A = C(X)$.

<u>Proof</u>: There are (distinct) points $x_1,\ldots,x_n$ such that each point x in the locally compact set $X_0 = X\setminus\{x_1,\ldots,x_n\}$ has a compact neighborhood of the indicated type. Let I be the ideal of functions in A which vanish on $\{x_1,\ldots,x_n\}$. Evidently I is uniformly closed. Let $B = I|X_0$. Evidently B is a uniformly closed subalgebra of $C_0(X_0)$. If $x,y \in X_0$ and $x \neq y$, set $x_0 = x$ and $x_{n+1} = y$. Use the fact that

$1 \in A$ and A separates points to find functions $f_j \in A$ $(j = 1,2,\ldots,n+1)$ with $f_j(x) = 1$, $f_j(x_j) = 0$ and functions $g_j \in A$ $(j = 0,1,\ldots,n)$ with $g_j(y) = 1$, $g_j(x_j) = 0$. Let $f = f_1 \ldots f_{n+1}$, $g = g_0 \ldots g_n$. Then for any $\alpha,\beta \in \mathbb{C}$ the function $\alpha f + \beta g$ belongs to I and has value α at x and β at y. Thus B separates the points of X_0.

We will show that each $x \in X_0$ has a compact neighborhood K_x such that $B|K_x = C(K_x)$. Consider the F_x provided by the hypothesis and use the result of the first paragraph to find a function $f_x \in I$ with $|f_x(x)| > 0$. Let K_x be a compact neighborhood of x inside F_x on which $|f_x| > 0$. (Note that then $K_x \subset X_0$.) If g belongs to $C(K_x)$, so does $g/f_x|K_x$. Let (Tietze) h be an extension of this function to F_x. There is then an $f \in A$ such that $f|F_x = h$. Then $f|K_x = g/f_x|K_x$ and so $ff_x|K_x = g$. Moreover $ff_x \in I$. It follows that $I|K_x = C(K_x)$ or, since $K_x \subset X_0$, $I|X_0|K_x = B|K_x = C(K_x)$.

We now apply Corollary 2.12 to assert that $B = C_0(X_0)$. It follows at once that I contains all continuous functions which vanish on $\{x_1,\ldots,x_n\}$. For each $j = 1,2,\ldots,n$ there is, by a construction like that in the first paragraph, a function $f_j \in A$ with $f_j(x_k) = \delta_{jk}$ $(j,k = 1,2,\ldots,n)$. Then for any $f \in C(X)$, the function $f - \sum_{j=1}^{n} f(x_j)f_j$ vanishes on $\{x_1,\ldots,x_n\}$, hence belongs to $I \subset A$. It follows that $f = \sum_{j=1}^{n} f(x_j)f_j + [f - \sum_{j=1}^{n} f(x_j)f_j]$ belongs to A.

Chapter III

WERMER'S THEOREM ON ALGEBRAS WITH MULTIPLICATIVELY CLOSED REAL PART

The principal result of this chapter, Corollary 3.6, is an easy consequence of the more general results of the next chapter, but we offer here the original proof of it because it is such an architectonic display of the tools of the trade in operation.

Theorem 3.1

Let X be a compact Hausdorff space, A a uniformly closed subalgebra of $C(X)$ which separates the points of X and contains the constants. Suppose also that A is an anti-symmetric algebra (i.e. contains no non-constant real functions) and that $\mathrm{Re}\, A$ is closed under multiplication. Then X is a single point.

Preliminaries

In lemmas 3.2, 3.3 and 3.5 to follow the notation and hypotheses will be that of theorem 3.1. Fix $x_0 \in X$. If $u \in \mathrm{Re}\, A$, the difference of any two functions in A with real part u is a pure imaginary-valued function, hence is constant by the antisymmetry of A. Therefore, recalling that $1 \in A$, there is a unique $f \in A$ satisfying

$$\mathrm{Re}\, f = u, \quad \mathrm{Im}\, f(x_0) = 0.$$

We denote this f by $H(u)$:

(1) $H(u) \in A$, $\operatorname{Re} H(u) = u$, $[\operatorname{Im} H(u)](x_0) = 0$.

Define a norm in $\operatorname{Re} A$ by

$$\|u\| = \|H(u)\|_\infty.$$

This is evidently a real linear space norm. We show in the next lemma that it is (essentially) an algebra norm. Since $H(\operatorname{Re} A)$ is the uniformly closed real subspace of all $f \in A$ such that $\operatorname{Im} f(x_0) = 0$, H effects an isometry of $(\operatorname{Re} A, \|\ \|)$ onto a complete space, so $\|\ \|$ is a complete norm.

<u>Lemma 3.2</u>

There exists a constant K such that

$$\|uv\| \leq K\|u\| \cdot \|v\| \qquad \forall u,v \in \operatorname{Re} A.$$

<u>Proof</u>: We first show that for each fixed $v \in \operatorname{Re} A$ the map $u \to uv$ is continuous in $\operatorname{Re} A$ by the (real) Closed Graph Theorem. Indeed suppose $\|u_n - u\| \to 0$ and $\|u_n v - w\| \to 0$ for some $u_n, u, w \in \operatorname{Re} A$. This means that for $f_n = H(u_n)$, $f = H(u)$, $g_n = H(u_n v)$, $g = H(w)$ we have

$$\|f_n - f\|_\infty \to 0 \quad \text{and} \quad \|g_n - g\|_\infty \to 0.$$

Therefore for every $x \in X$

$$|f_n(x) - f(x)| \to 0 \quad \text{and} \quad |g_n(x) - g(x)| \to 0$$

whence

$$u_n(x) = \operatorname{Re} f_n(x) \to \operatorname{Re} f(x) = u(x) \quad \text{and}$$

$$(u_n v)(x) = \operatorname{Re} g_n(x) \to \operatorname{Re} g(x) = w(x)$$

whence

$$(u_n v)(x) = u_n(x)v(x) \to u(x)v(x)$$

and so

$$u(x)v(x) = w(x),$$

that is, $uv = w$. Hence the graph of $u \to uv$ is closed and so the map is continuous. Let $K(v)$ be its norm. Then for fixed u the family

$$\{uv : v \in \text{Re } A, \|v\| \le 1\}$$

is bounded:

$$\|uv\| \le \|v\| K(u) \le K(u).$$

Hence by the (real) Uniform Boundedness Principle the family of bounds $\{K(v): v \in \text{Re } A, \|v\| \le 1\}$ is bounded, say by K. Then for any $u,v \in \text{Re } A$ with $\|u\| \le 1$ and $\|v\| \le 1$ we have

$$\|uv\| \le \|u\| K(v) \le K(v) \le K.$$

So $\quad \|uv\| \le K\|u\| \, \|v\| \qquad \forall u,v \in \text{Re } A.$

Next define B to be the set of all functions $u + iv$ with $u,v \in \text{Re } A$. Then B is a complex subalgebra (since $\text{Re } A$ is closed under multiplication) of $C(X)$ which is conjugate closed and contains A. Define

$$\|u + iv\| = \|u\| + \|v\| \qquad \forall u,v \in \text{Re } A$$

$$\|f\|_B = \sup_\theta \|e^{i\theta} f\| \qquad \forall f \in B.$$

Evidently $\| \;\|_B$ is a complex linear space norm on B. Notice that if $f = u + iv$ then for any $e^{i\theta} = a + ib$ we have

$$\|e^{i\theta} f\| = \|(au-bv) + i(av+bu)\| = \|au-bv\| + \|av+bu\|$$

$$\le |a|\|u\| + |b|\|v\| + |a|\|v\| + |b|\|u\| = (|a|+|b|)(\|u\|+\|v\|)$$

$$= (|a| + |b|)\|f\| \le 2\|f\|.$$

Taking the supremum over all θ, we find

$$\|f\|_B \le 2\|f\|.$$

<u>Lemma 3.3</u>

B is complete in $\| \|_B$ and this norm satisfies

$$\|f\,g\|_B \le K\|f\|_B\|g\|_B \qquad \forall f,g \in B.$$

Hence B is a complex Banach algebra under a norm equivalent to $\| \|_B$.

<u>Proof</u>: If $f_n \in B$ and $\|f_n - f_m\|_B \to 0$ then, since $\| \|_B \ge \| \|$, we have $\|f_n - f_m\| = \|u_n - u_m\| + \|v_n - v_m\| \to 0$, where $u_n = \text{Re } f_n$, $v_n = \text{Im } f_n$. As Re A is complete in $\| \|$, there exist $u,v \in \text{Re } A$ with $\|u_n - u\| \to 0$ and $\|v_n - v\| \to 0$. Let $f = u + iv \in B$. Then

$$\|f_n - f\|_B \le 2\|f_n - f\| = 2\|u_n - u\| + 2\|v_n - v\| \to 0.$$

Finally for $f,g \in B$ with, say, $u = \text{Re } f$, $v = \text{Im } f$, $u' = \text{Re } g$, $v' = \text{Im } g$, we have

$$\|fg\| = \|(uu' - vv') + i(uv' + u'v)\| = \|uu' - vv'\| + \|uv' + u'v\|$$

$$\le K\|u\|\,\|u'\| + K\|v\|\,\|v'\| + K\|u\|\,\|v'\| + K\|u'\|\,\|v\|$$

$$= K(\|u\|+\|v\|)(\|u'\|+\|v'\|) = K\|f\|\,\|g\|.$$

Therefore for all real θ

$$\|e^{i\theta}fg\| \le K\|e^{i\theta}f\|\,\|g\| \le K\|f\|_B\|g\|_B$$

and taking the supremum on θ

$$\|fg\|_B \le K\|f\|_B\|g\|_B.$$

Next we present a working lemma about analytic functions which we need here and also later on.

<u>Lemma 3.4</u>

Given $\lambda > 0$ there exist a polynomial p and an $\epsilon > 0$

such that p maps the open disk $\{z \in \mathbb{C}: |z| < 1+\varepsilon\}$ into the corridor $|\mathrm{Re}\ z| < 1$ and satisfies $p(0) = 0$, $\mathrm{Im}\ p(1) > \lambda$.

<u>Proof</u>: For each $r > 0$ let $D_r = \{z \in \mathbb{C}: |z| < r\}$. First note that the series

$$(1) \qquad \ell(z) = -\sum_{n=1}^{\infty} \frac{1}{n} z^n$$

converges absolutely in D_1 and so defines an analytic function there which may be differentiated termwise to give

$$(2) \qquad \ell'(z) = -\sum_{n=1}^{\infty} z^{n-1} = -(1-z)^{-1}.$$

It follows that the analytic function $(1-z)^{-1}e^{\ell(z)}$ in D_1 has derivative $\dfrac{(1-z)[e^{\ell(z)}]' + e^{\ell(z)}}{(1-z)^2} = \dfrac{(1-z)\ell'(z)e^{\ell(z)} + e^{\ell(z)}}{(1-z)^2} = 0$

and so this function is constant in the disk D_1:

$$(1-z)^{-1}e^{\ell(z)} \equiv (1-0)^{-1}e^{\ell(0)} = 1$$

$$e^{\ell(z)} = 1 - z \qquad \forall z \in D_1.$$

It follows that $\cos(\mathrm{Im}\ \ell(z)) = \mathrm{Re}[\dfrac{1-z}{|1-z|}] = \dfrac{1}{|1-z|}(1-\mathrm{Re}\ z) > 0$

and therefore $\mathrm{Im}\ \ell(z) \in 2\pi\mathbb{Z} + (-\pi/2,\pi/2)$. But $\mathrm{Im}\ \ell(D_1)$ is the image of the connected set D_1 under the continuous map $\mathrm{Im}\ \ell$ and so is a connected set. This implies that $\mathrm{Im}\ \ell(D_1) \subset 2\pi n + (-\pi/2,\pi/2)$ for a single integer n. And $\ell(0) = 0$ shows that $n = 0$:

$$(4) \qquad \mathrm{Im}\ \ell(D_1) \subset (-\pi/2,\pi/2).$$

Next we choose $r_3 > r_2 > r_1 > 1$ so that

$$(5) \qquad \log[\frac{1+r_3^{-1}}{1-r_3^{-1}}] > \frac{\pi}{2}\lambda.$$

Now for $|w| < 1$ we have $\text{Re}[\frac{1+w}{1-w}] = \frac{1}{2}[\frac{1+w}{1-w} + \frac{1+\bar{w}}{1-\bar{w}}] = \frac{1-|w|^2}{|1-w|^2} > 0$. Therefore $\{(1 + r_3^{-1}z)(1 - r_3^{-1}z)^{-1}: z \in \bar{D}_{r_2}\}$ is a compact subset of the open right half plane and so for some $t > 1$ lies in $\{w \in \mathbb{C}: |w-t| < t\}$. We can therefore form

(6) $$f(z) = \frac{2i}{\pi}[\ell(1 - \frac{1}{t} \cdot \frac{1+r_3^{-1}z}{1-r_3^{-1}z}) + \log t], \quad z \in \bar{D}_{r_2}.$$

We have from (4) and (6) that the compact set $\text{Re}\, f(\bar{D}_{r_2})$ lies in $(-1,1)$ and so

(7) $$\sup|\text{Re}\, f(\bar{D}_{r_2})| = \eta < 1.$$

Also from (6)

$$\text{Im}\, f(1) = \frac{2}{\pi}[\text{Re}\, \ell(1 - \frac{1}{t} \cdot \frac{1 + r_3^{-1}}{1 - r_3^{-1}}) + \log t]$$

$$= \frac{2}{\pi}[\log(\frac{1}{t} \cdot \frac{1 + r_3^{-1}}{1 - r_3^{-1}}) + \log t] \qquad \text{by (3)}$$

(8) $$= \frac{2}{\pi} \log[\frac{1 + r_3^{-1}}{1 - r_3^{-1}}] > \lambda \qquad \text{by (5)}.$$

We set

(9) $$\delta = \text{Im}\, f(1) - \lambda > 0,$$

and then choose a partial sum p in the power series expansion of f about 0 in D_{r_2} such that in the compact subset $\bar{D}_{r_1}$ we have

(10) $$|p(z) - f(z)| < \min\{\delta, 1-\eta\} \qquad \forall z \in \bar{D}_{r_1}.$$

Then

(11) $$p(0) = f(0) = 0 \qquad \text{by (6) and (3)}$$

and for all $z \in \bar{D}_{r_1}$

$$(12)\quad |\mathrm{Re}\ p(z)| \le |\mathrm{Re}\ f(z)|+|p(z)-f(z)| \overset{(10)}{<} |\mathrm{Re}\ f(z)|+1-\eta \overset{(7)}{\le} \eta+(1-\eta)=1$$

$$(13)\quad \mathrm{Im}\ p(1) \ge \mathrm{Im}\ f(1)-|\mathrm{Im}\ p(1)-\mathrm{Im}\ f(1)| \ge \mathrm{Im}\ f(1)-|p(1)-f(1)|$$

$$\overset{(10)}{>} \mathrm{Im}\ f(1) - \delta \overset{(9)}{=} \lambda.$$

Lemma 3.5

If $u \in \mathrm{Re}\ A$ and $u > 0$ on X then $\log\circ u \in \mathrm{Re}\ A$.

Proof: By compactness $u \ge \delta > 0$ on X for some $\delta > 0$. If φ is any non-zero complex homomorphism of A, then φ is a continuous functional of norm 1 (see III, p. 1) and so lifts to a functional of norm 1 on $C(X)$ by the Hahn-Banach Theorem. This extension is represented by a (complex, regular, Borel) measure μ on X which satisfies $\|\mu\| = 1 = \mu(1)$ and so is a positive measure. Then if $f \in A$ is such that $\mathrm{Re}\ f = u$

$$(*)\quad \mathrm{Re}\ \varphi(f) = \mathrm{Re} \int_X f d\mu = \int_X \mathrm{Re}\ f d\mu = \int_X u d\mu \ge \int_X \delta d\mu = \delta > 0.$$

Now if ψ is a non-zero complex homomorphism of B then $g \to \psi(g)$ and $g \to \overline{\psi(\overline{g})}$ are non-zero complex homomorphisms on the subalgebra A of B. Call them φ_1, φ_2 respectively. With f as before $\psi(u) = \psi(\frac{1}{2}(f+\overline{f})) = \frac{1}{2}\psi(f) + \frac{1}{2}\psi(\overline{f}) = \frac{1}{2}\varphi_1(f) + \frac{1}{2}\overline{\varphi_2(f)}$ and therefore $\mathrm{Re}\ \psi(u) = \frac{1}{2}\mathrm{Re}\ \varphi_1(f) + \frac{1}{2}\mathrm{Re}\ \varphi_2(f) > 0$ by two applications of (*). This says that the spectrum of u in the Banach algebra B lies in the open right half plane. As the spectrum is compact it therefore lies in $\{z \in \mathbb{C}: |z - r| < r\}$ for some $r > 0$. Consider $v = 1 - r^{-1}u$. Its B-spectrum lies in the open unit disk so its B-spectral radius is less than 1; that is, by the Gelfand spectral radius formula, $\lim_{n\to\infty}\|v^n\|_B^{1/n} = \rho < 1$. Therefore by the root test $\sum_{n=1}^{\infty} n^{-1}\|v^n\|_B$ converges and so the series $-\sum_{n=1}^{\infty} n^{-1}v^n$ converges in B, say to w. Now $\|\ \|_\infty \le \|\ \|_B$ so for each $x \in X$ evaluation at x is continuous in B and we get $w(x) = -\sum_{n=1}^{\infty} n^{-1}v^n(x) = -\sum_{n=1}^{\infty} n^{-1}(v(x))^n$ and this numerical series converges (recall $|v(x)| \le \rho < 1$) to $\log(1-v(x)) = \log(r^{-1}u(x))$ [recall equations (1) and (3) in the proof of Lemma 3.4]. Therefore $\log\circ(1-v) = w$. In particular $w \in \operatorname{Re} B = \operatorname{Re} A$ and finally $\log\circ u = \log\circ(1-v) + \log r = w + \log r \in \operatorname{Re} A$.

<u>Proof of Theorem 3.1</u>: We suppose $X \neq \{x_0\}$ and deduce a contradiction. Then there exists $g \in A$ with $g(x_0) = 0$ and $\|g\|_\infty = 1$. Let $a \in X$ and $|g(a)| = 1$. The magic number is $\lambda = 16K + 5$, where K is the constant of lemma 3.2. Let p be as in Lemma 3.4 for this λ and define $q(z) = \frac{1}{2} + \frac{1}{2}p(\overline{g(a)}z)$. Then q maps the disk $|z| \le 1$ into the corridor $0 < \operatorname{Re} z < 1$. Moreover $\operatorname{Im} q(0) = \frac{1}{2}\operatorname{Im} p(0) = 0$ and $\operatorname{Im} q(g(a)) = \frac{1}{2}\operatorname{Im} p(1) > \frac{1}{2}\lambda$. And as q is a polynomial, $q\circ g$ belongs to A. Call it f. Thus

(1) $\quad 0 < \operatorname{Re} f < 1 \quad$ on X

(2) $\quad \operatorname{Im} f(x_0) = 0$

(3) $\quad \|f\|_\infty \ge \operatorname{Im} f(a) = \operatorname{Im} q(g(a)) > \frac{1}{2}\lambda$.

Now by lemma 3.5, $\log\circ\operatorname{Re} f \in \operatorname{Re} A$, i.e. there exists $F \in A$ with

$$\operatorname{Re} F = \log\circ\operatorname{Re} f.$$

Set $V = \exp\circ\frac{1}{2}F \in A$ and get

(4) $|V|^2 = \text{Re } f$

(5) $\|V\|_\infty = \|\text{Re } f\|_\infty^{\frac{1}{2}} < 1$ by (1).

Now for any complex number z we have

$$(\text{Re } z)^2 = \tfrac{1}{2}(\text{Re } z^2 + |z|^2)$$

and applying this to various $V(x)$ we get

(6) $(\text{Re } V)^2 = \frac{1}{2}(\text{Re } V^2 + |V|^2) = \frac{1}{2}\text{Re}(V^2 + f)$ by (4).

Observe that for any $h \in A$ we clearly have that $H(\text{Re } h) = h - i \text{ Im } h(x_0)$ and so

$$\|\text{Re } h\| = \|H(\text{Re } h)\|_\infty \geq \|h\|_\infty - |\text{Im } h(x_0)|.$$

Therefore from (6)

$$\|(\text{Re } V)^2\| = \tfrac{1}{2}\|\text{Re}(V^2+f)\| \geq \tfrac{1}{2}[\|V^2+f\|_\infty - |\text{Im}(V^2(x_0)+f(x_0))|]$$

$$\geq \tfrac{1}{2}[\|f\|_\infty - \|V\|_\infty^2 - |\text{Im } V^2(x_0)|] \qquad \text{by (2)}$$

(7) $\geq \frac{1}{2}[\frac{1}{2}\lambda - 2]$ by (3) and (5).

On the other hand by lemma 3.2

(8) $\|(\text{Re} V)^2\| \leq K\|\text{Re } V\|^2 \leq K[2\|V\|_\infty]^2$

since for any $h \in A$, $H(\text{Re } h) = h - i \text{ Im } h(x_0)$ and so $\|\text{Re } h\| = \|H(\text{Re } h)\|_\infty \leq 2\|h\|_\infty$. From (8) and (5) we get

$$\|(\text{Re } V)^2\| \leq 4K$$

which with (7) leads to

$$4K \geq \tfrac{1}{4}\lambda - 1$$

$$16K + 4 \geq \lambda,$$

manifestly contrary to the definition of λ.

<u>Corollary 3.6</u>

(Wermer [50]) Let X be a compact Hausdorff space, A

a uniformly closed subalgebra of $C(X)$ which separates the points of X and contains the constants. If $\operatorname{Re} A$ is a ring, then $A = C(X)$.

Proof: Let $\mathcal{K}$ be the maximal A-antisymmetric decomposition of X. Then $A|K$ is a closed subalgebra of $C(K)$ for each $K \in \mathcal{K}$ by Bishop's Theorem (1.2) and evidently $\operatorname{Re}(A|K) = \operatorname{Re} A|K$ is a ring. Moreover, $A|K$ contains no non-constant real functions. By Theorem 3.1 each K is a single point. But then $f|K \in A|K$ for each $K \in \mathcal{K}$ and every $f \in C(X)$ and so by Bishop's Theorem $f \in A$ for every $f \in C(X)$.

Corollary 3.7

(Kahane 1961, unpublished) There exists a continuous real function u on the unit circle having a continuous real conjugate v, in the sense that $u + iv$ extends analytically into the disk, but such that u^2 does not have such a conjugate.

Proof: For A in Corollary 3.6 take the disk algebra: continuous complex functions on the circle admitting analytic extensions into the disk. $\operatorname{Re} A$ is then the class of real continuous functions having real continuous conjugates. Evidently A does not comprise all the continuous functions on the unit circle. For example, $f(z) = \overline{z}$ does not belong to A: if F were an analytic extension of f then $zF(z) = z\overline{z} = 1$ would hold on the boundary of the unit disk, hence throughout the disk by the Uniqueness Theorem for Analytic Functions. But the validity of this equality for $z = 0$ is a manifest absurdity. By Corollary 3.6 then $\operatorname{Re} A$

is not closed under multiplication, hence not closed under squaring either because of the identity $2ab = (a+b)^2 - a^2 - b^2$.

Remarks

Harmonic analyzers may wish to translate the corollary into the language of conjugate functions in the sense of Fourier Theory. We also remark that it is easy to adapt the techniques of lemma 3.2 to prove some interesting relatives of the last corollary. We refer the reader to Brown [12] for these.

Chapter IV

THE WORK OF ALAIN BERNARD

In this chapter we present some exciting recent work of Alain Bernard. This work is scattered among Comptes Rendus notes [4], [5], [6] but will probably be given a unified exposition in [7]. The account here follows some Grenoble seminar notes of Bernard and the recent lectures [22] of Glicksberg. We thank both of them for making this material available to us.

<u>Lemma 4.1</u>

Let X be a compact Hausdorff space, A a subalgebra of $C(X)$ which contains the constants and separates the points of X, K a maximal set of antisymmetry of $A \cap C_{\mathbb{R}}(X)$, $\lambda > 0$. If K is not a singleton, there exist $x_1, x_2 \in K$, $f \in A$ such that $f(x_1) = 0$, $\text{Im}\, f(x_2) > \lambda$, $\|\text{Re}\, f\|_\infty \le 1$.

<u>Proof</u>: There is $g \in A$ which is non-constant on K. Pulling off an appropriate constant and taking an appropriate scalar multiple, we can have

$$g(x_1) = 0 \qquad \text{some } x_1 \in K$$

$$1 = g(x_2) = \sup\{|g(x)|: x \in K\}, \quad \text{some } x_2 \in K.$$

By lemma 3.4 there exists a polynomial p and an $\epsilon > 0$ such that p maps the open disk $D = \{z \in \mathbb{C}: |z| < 1 + \epsilon\}$ into the corridor $|\text{Re}\, z| < 1$ and satisfies $p(0) = 0$, $\text{Im}\, p(1) > \lambda$.

Set $H = \{x \in X: |g(x)| \geq 1 + \epsilon\}$. This set is compact and disjoint from K. Because K is a maximal set of antisymmetry of $A \cap C_{\mathbb{R}}(X)$, there exists for each $x \notin K$ a function $h_x \in A$ which is real on all of X but not constant on $\{x\} \cup K$ (though of course constant on K); say, without loss of generality (recalling that $\|g\|_\infty \geq |g(x_2)| = 1$)

$$h_x(K) = 1, \ |h_x(x)| < \frac{1}{\|g\|_\infty}, \ \|h_x\|_\infty \leq 1.$$

For example, if $\varphi \in A \cap C_{\mathbb{R}}(X)$ and $\varphi(K) = c \neq \varphi(x)$ consider first $\psi = [\frac{\varphi - c}{\|\varphi - c\|_\infty}]^2 \in A \cap C_{\mathbb{R}}(X)$. Then $0 \leq \psi \leq 1$, $\psi(K) = 0$, $\psi(x) \neq 0$. For h_x take a sufficiently high power of $1 - \psi$. There is a neighborhood N_x of x in which the middle inequality above persists:

$$\sup\{|h_x(y)|: y \in N_x\} < \frac{1}{\|g\|_\infty}.$$

Cover compact H with $N_{x_3}, N_{x_4}, \ldots, N_{x_n}$ $(x_3, x_4, \ldots, x_n \in H)$ and set $h = h_{x_3} \cdot h_{x_4} \cdot \ldots \cdot h_{x_n}$ and get

$$h(K) = 1, \ \sup\{|h(x)|: x \in H\} \leq \frac{1}{\|g\|_\infty}.$$

For the function $g' = g \cdot h$ in A we have $g'(x_1) = 0$, $g'(x_2) = 1$,

$$|g'(x)| = |g(x)|\,|h(x)| \leq \begin{cases} |g(x)| < 1 + \epsilon & x \notin H \\ \|g\|_\infty \cdot \frac{1}{\|g\|_\infty} = 1 & x \in H \end{cases}$$

and so since X is compact

$$\|g'\|_\infty < 1 + \epsilon.$$

The function $f = p \circ g'$ in A then satisfies

$$f(x_1) = p(g'(x_1)) = p(0) = 0$$
$$\operatorname{Im} f(x_2) = \operatorname{Im} p(g'(x_2)) = \operatorname{Im} p(1) > \lambda$$
$$|\operatorname{Re} f(x)| = |\operatorname{Re} p(g'(x))| < 1 \qquad \text{for } x \in X,$$

and so $\|\operatorname{Re} f\|_\infty < 1$.

Lemma 4.2

Let X be a compact Hausdorff space, A a Banach algebra lying in $C(X)$ which separates the points of X and contains the constants. If $\operatorname{Re} A$ is uniformly closed in $C_{\mathbb{R}}(X)$, then

$$A_{\mathbb{R}} = A \cap C_{\mathbb{R}}(X)$$

separates the points of X.

Proof: Let $\| \ \|_A$ denote the norm in A. Recall (III, p. 1) that for all $f \in A$

$$\|f\|_\infty \le \|f\|_A.$$

Therefore

$$f \to \operatorname{Re} f$$

is a norm decreasing map of A onto $\operatorname{Re} A$, the latter in its uniform norm. As $\operatorname{Re} A$ is by hypothesis a Banach space in this norm, the Open Map Theorem (for real Banach spaces) provides an $M > 0$ such that

(1) $u \in \operatorname{Re} A \;\&\; \|u\|_\infty \le 1 \Rightarrow \exists v \in C_{\mathbb{R}}(X)$ such that $u+iv \in A \;\&$ $\|u+iv\|_A < M$.

If we suppose, contrary to what is claimed, that $A_{\mathbb{R}}$ does not separate the points of X, then some (maximal) $A_{\mathbb{R}}$-antisymmetry set K is non-degenerate and lemma 4.1 then provides $x_1, x_2 \in K$ and f such that

(2) $f \in A,\ f(x_1) = 0,\ \text{Im } f(x_2) > 2M + 1,\ \|\text{Re } f\|_\infty \le 1.$

Apply (1) to $u = \text{Re } f$. With this v consider $h = \frac{f - (u+iv)}{i}$. Evidently $h \in A$ and h is real-valued on X. But $f(x_1) = 0$ so $u(x_1) = \text{Re } f(x_1) = 0$ and $h(x_1) = -v(x_1)$,

$$|h(x_1)| = |v(x_1)| \le \|v\|_\infty \le \|u + iv\|_\infty \le \|u + iv\|_A \le M$$

$$\begin{aligned}|h(x_2)| &= |i \text{ Im } f(x_2) - iv(x_2)| \ge |\text{Im } f(x_2)| - |v(x_2)| \ge 2M+1 - |v(x_2)| \\ &\ge 2M+1 - \|v\|_\infty \ge 2M+1 - \|u + iv\|_\infty \ge 2M+1 - \|u + iv\|_A \\ &\ge M + 1 \qquad \text{by (1).}\end{aligned}$$

Therefore $h \in A \cap C_{\mathbb{R}}(X)$ but $h(x_1) \ne h(x_2)$, contrary to the fact that x_1, x_2 both belong to the set K which is a set of antisymmetry for $A \cap C_{\mathbb{R}}(X)$.

<u>Corollary 4.3</u>

(Hoffman & Wermer [29]) If X is a compact Hausdorff space and A is a uniformly closed subalgebra of $C(X)$ which separates points and contains the constants, then $\text{Re } A$ uniformly closed implies $A = C(X)$.

<u>Proof</u>: The last lemma allows real Stone-Weierstrass to be applied to the real algebra $A_{\mathbb{R}} = A \cap C_{\mathbb{R}}(X)$. Therefore $A_{\mathbb{R}}$ is uniformly dense in $C_{\mathbb{R}}(X)$. But this real subalgebra of $C_{\mathbb{R}}(X)$ is uniformly closed since A is uniformly closed in $C(X)$, and so $A_{\mathbb{R}} = C_{\mathbb{R}}(X)$. Hence $A \supset A_{\mathbb{R}} + iA_{\mathbb{R}} = C_{\mathbb{R}}(X) + iC_{\mathbb{R}}(X) = C(X)$.

<u>Note</u>

A. Browder gives another nice proof of this on pp. 88-89 of [8]. Also Arenson [1] gives a proof.

More generally

Corollary 4.4

If X is a compact Hausdorff space, A a Banach algebra lying in $C(X)$ which contains the constants and separates the points of X, then $\text{Re } A$ uniformly closed in $C_{\mathbb{R}}(X)$ implies $\text{Re } A = C_{\mathbb{R}}(X)$.

Proof: By lemma 4.2 and real Stone-Weierstrass, the real algebra $A_{\mathbb{R}}$ is uniformly dense in $C_{\mathbb{R}}(X)$. Since $\text{Re } A \supset A_{\mathbb{R}}$ and $\text{Re } A$ is assumed to be uniformly closed, we must have $\text{Re } A = C_{\mathbb{R}}(X)$.

Lemma 4.5

(Bernard's Lemma) Suppose E,F are (real or complex) normed linear spaces, $E \subset F$ and the injection is continuous. Let $\tilde{E} = \ell_\infty(\mathbb{N},E)$, the bounded functions from $\mathbb{N} = \{1,2,....\}$ to E normed as usual, and $\tilde{F} = \ell_\infty(\mathbb{N},F)$. Then $\tilde{E} \subset \tilde{F}$. If in addition E is complete and $\tilde{E}$ is dense in $\tilde{F}$ then $E = F$.

Proof: By hypothesis there is a constant M such that

(1) $$\| \ \|_F \le M \| \ \|_E.$$

Therefore any bounded sequence in E is bounded in F so $\tilde{E} \subset \tilde{F}$ and evidently

$$\| \ \|_{\tilde{F}} \le M \| \ \|_{\tilde{E}}.$$

Now suppose E is complete and $\tilde{E}$ is dense in $\tilde{F}$. We will show first that for some $r > 0$

(2) $$x \in F \ \& \ \|x\|_F < 1 \Rightarrow \exists y \in E, \ \|y\|_E < r \ \& \ \|x-y\|_F < 1/2.$$

Indeed, if this fails for every $r > 0$, we can select a

sequence of points $x_n \in F$ with

(3) $\|x_n\|_F < 1$

such that

(4.n) $\|x_n - y\|_F \geq \frac{1}{2}$ for every $y \in E$ with $\|y\|_E < n$.

Now $\{x_n\} \in \tilde{F}$ and $\tilde{E}$ is dense therein, so there exists $\{y_n\} \in \tilde{E}$ with $\|\{x_n\} - \{y_n\}\|_{\tilde{F}} \leq \frac{1}{4}$, that is,

(5) $\|x_n - y_n\|_F \leq \frac{1}{4}$ for each $n = 1,2,\ldots$.

But $\{y_n\} \in \tilde{E}$ means that for some constant K

(6) $\|y_n\|_E \leq K$ $n = 1,2,\ldots$.

Then (5) and (6) contradict (4.n) when $n > K$.

Finally we use (2) to prove that $F \subset E$. It suffices to show that each $x \in F$ with $\|x\|_F < 1$ belongs to E. Let $y = y_1$ be chosen according to (2). Suppose $y_1,\ldots,y_n \in E$ chosen so that

(7.n) $\|x - (y_1 + \ldots + y_n)\|_F \leq 2^{-n}$

(8.n) $\|y_n\|_E \leq r2^{-n+1}$.

Apply (2) (appropriately scaled) to $x - (y_1 + \ldots + y_n)$ to find $y_{n+1} \in E$ with

(8.n+1) $\|y_{n+1}\|_E \leq r\|x - (y_1 + \ldots + y_n)\|_F$

$\leq r2^{-n}$ by (7.n)

such that

(7.n+1) $\|x - (y_1 + \ldots + y_n) - y_{n+1}\|_F \leq \frac{1}{2}\|x-(y_1+\ldots+y_n)\|_F$

$\leq 2^{-n-1}$ by (7.n).

This completes the inductive construction. Since E is

complete, the series $\sum_{n=1}^{\infty} y_n$ converges in E by (8.n), say to y. Therefore by (1) the series also converges to y in F. But by (7.n) the series converges to x in F. It follows that $x = y \in E$.

Theorem 4.6

(Bernard) Let X be a compact Hausdorff space, A a Banach algebra lying in $C(X)$ which separates points and contains the constants. If $\operatorname{Re} A$ is uniformly closed, then $A = C(X)$.

Proof: Let bar denote uniform closure. Since $\operatorname{Re} A$ is uniformly closed, $\operatorname{Re} \overline{A} \subset \overline{\operatorname{Re} A} = \operatorname{Re} A$ so $\operatorname{Re} \overline{A} = \operatorname{Re} A$ is closed. The Hoffman-Wermer result (4.3) is therefore applicable to $\overline{A}$ and we conclude $\overline{A} = C(X)$. Whence

(1) $\operatorname{Re} A = \operatorname{Re} \overline{A} = C_{\mathbb{R}}(X)$.

Now, as observed before (III, p. 1), we have $\| \ \|_X \le \| \ \|_A$ and therefore if we define

$$\|u\| = \inf\{\|f\|_A : f \in A, \operatorname{Re} f = u\} \qquad u \in \operatorname{Re} A = C_{\mathbb{R}}(X)$$

then $\| \ \|$ is a complete (quotient) norm in $C_{\mathbb{R}}(X)$ which dominates the uniform norm. By the Open Map Theorem there is a constant K such that $\| \ \| \le K\| \ \|_X$. Since $\| \ \|$ is a quotient norm, this implies

(2) $\forall u \in C_{\mathbb{R}}(X)\ \exists v$ such that $u + iv \in A$ and $\|u + iv\|_A \le (K + 1)\|u\|_X$.

Since $\| \ \|_X \le \| \ \|_A$

(3) $\widetilde{A} = \ell_\infty(\mathbb{N}, A)$ is a subalgebra of $\ell_\infty(\mathbb{N}, C(X)) = \widetilde{C}(X)$

whence

$$\operatorname{Re} \tilde{A} \subset \operatorname{Re} \ell_\infty(\mathbb{N}, C(X)).$$

But (2) implies that $\operatorname{Re} \ell_\infty(\mathbb{N}, C(X)) \subset \operatorname{Re} \tilde{A}$, for if $\{u_n\} \in \operatorname{Re} \ell_\infty(\mathbb{N}, C(X))$ [$= \ell_\infty(\mathbb{N}, C_{\mathbb{R}}(X))$ note], say $\|u_n\|_X \leq M$ for all n, then by (2) there exist v_n such that $u_n + iv_n \in A$ and $\|u_n + iv_n\|_A \leq (K + 1)M$ and so $\{u_n + iv_n\} \in \tilde{A}$. We have therefore

(4) $$\operatorname{Re} \tilde{A} = \operatorname{Re} \ell_\infty(\mathbb{N}, C(X)).$$

Now if $\overline{\mathbb{N} \times X}$ denotes the Stone-Čech compactification of $\mathbb{N} \times X$, then $\ell_\infty(\mathbb{N}, C(X))$ is naturally identified with $C(\overline{\mathbb{N} \times X})$ and $\tilde{A}$ with a subalgebra thereof and (4) translates into

$$\operatorname{Re} \tilde{A} = C_{\mathbb{R}}(\overline{\mathbb{N} \times X}).$$

In particular $\tilde{A}$ separates the points of $\overline{\mathbb{N} \times X}$ and, with bar denoting uniform closure in $C(\overline{\mathbb{N} \times X})$, $\operatorname{Re} \bar{\tilde{A}} \subset \overline{\operatorname{Re} \tilde{A}} \subset C_{\mathbb{R}}(\overline{\mathbb{N} \times X}) = \operatorname{Re} \tilde{A}$. Hoffman-Wermer (4.3) therefore applies to $\bar{\tilde{A}}$ and says that

$$\bar{\tilde{A}} = C(\overline{\mathbb{N} \times X}),$$

that is,

(6) $\tilde{A}$ is dense in $\ell_\infty(\mathbb{N}, C(X)) = \tilde{C}(X)$.

It follows from Lemma 4.5 that $A = C(X)$.

<u>Corollary 4.7</u>

(Sidney & Stout [47]) If X is a compact Hausdorff space, A is a uniformly closed point-separating subalgebra of $C(X)$ which contains the constants, Y is a closed subset of X and $\operatorname{Re} A|Y$ is uniformly closed in $C_{\mathbb{R}}(Y)$, then

$A|Y = C(Y)$.

Proof: Apply Theorem 4.6 to the algebra $A|Y$ with its quotient norm (IV, p. 2).

Our next corollary (and the proof given) is valid in any compact abelian group with totally ordered dual, as the reader will at once perceive. The idea of deducing this result from Bernard's theorem is due to J.-P. Kahane.

Corollary 4.8

(Wik [51]). For a closed subset E of the circle T, the following are equivalent

(i) $C(E) = \ell_1(Z)^\wedge|E$

(ii) $C(E) = \ell_1(Z^+)^\wedge|E$.

Proof: (ii) $\Rightarrow$ (i) is obvious since $\ell_1(Z^+)^\wedge|E \subset \ell_1(Z)^\wedge|E \subset C(E)$. For the converse assume that the closed set E satisfies (i). Let $A = \ell_1(Z^+)^\wedge|E$ with the quotient norm of $\ell_1(Z^+)/kE$ where $kE = \{\psi \in \ell_1(Z^+): \hat{\psi}(E) = 0\}$. Thus A is a complex Banach algebra lying in $C(E)$. Given $f \in C_{\mathbb{R}}(E)$, by (i) there is a $\varphi \in \ell_1(Z)$ such that

(1) $f = \hat{\varphi}|E$.

Defining $\psi \in \ell_1(Z)$ by

(2) $\psi(n) = \frac{1}{2}(\varphi(n) + \overline{\varphi(-n)})$

we have $\hat{\psi} = \frac{1}{2}(\hat{\varphi} + \overline{\hat{\varphi}}) = \operatorname{Re} \hat{\varphi}$ and therefore, as f is real-valued, (1) gives

(3) $f = \hat{\psi}|E$.

Since $\psi(-n) = \overline{\psi(n)}$ by (2), we have for every $z \in T$

$$\hat{\psi}(z) = \sum_{n=-\infty}^{\infty} \psi(n)z^n = \sum_{n=0}^{\infty} \psi(n)z^n + \sum_{n=1}^{\infty} \psi(-n)z^{-n} = \sum_{n=0}^{\infty} \psi(n)z^n + \sum_{n=1}^{\infty} \overline{\psi(n)}\overline{z}^n$$

$$= \sum_{n=0}^{\infty} \psi(n)z^n + \overline{\sum_{n=0}^{\infty} \psi(n)z^n} - \psi(0) = \mathrm{Re}[-\psi(0) + \sum_{n=0}^{\infty} 2\psi(n)z^n]$$

$$= \mathrm{Re}[\chi_{Z^+}\cdot 2\psi - \psi(0)\cdot\chi_{\{0\}}]^\wedge(z), \text{ that is,}$$

$$(4) \quad \hat{\psi} \in \mathrm{Re}[\ell_1(Z^+)^\wedge].$$

It follows from (3) and (4), since $f \in C_{\mathbb{R}}(E)$ is arbitrary, that

$$C_{\mathbb{R}}(E) = \mathrm{Re}\ A.$$

A fortiori then A separates the points of E and contains the constants and $\mathrm{Re}\ A$ is closed in $C_{\mathbb{R}}(E)$. So by Bernard's Theorem (4.6) $A = C(E)$.

Definition 4.9

Let X be a compact Hausdorff space, $A \subset C(X)$ a (real or complex) normed linear space continuously injected in $C(X)$. A is called ultraseparating if $\tilde{A} = \ell_\infty(\mathbb{N},A)$, regarded [cf. lemma 4.5] as a subset of $\ell_\infty(\mathbb{N},C(X)) = C(\overline{\mathbb{N}\times X})$, separates the points of $\overline{\mathbb{N}\times X}$, the latter space being the Stone-Čech compactification of $\mathbb{N}\times X$.

Lemma 4.10

Let X be a compact Hausdorff space, $A \subset C(X)$ a (real or complex) normed linear space continuously injected in $C(X)$. If A is ultraseparating then for each compact $K \subset X$ the space $A|K$ with the quotient norm is continuously injected in $C(K)$ and is ultraseparating.

<u>Proof</u>: By hypothesis there is a constant M such that $\| \ \|_X \leq M \| \ \|_A$. Thus for any $g \in A|K$ and any $f \in A$ with $f|K = g$ we have

$$\|g\|_K \leq \|f\|_X \leq M\|f\|_A.$$

Therefore

$$\|g\|_K \leq M \inf\{\|f\|_A : f \in A,\ f|K = g\}.$$

As the latter infimum is the definition of $\|g\|_{A|K}$, we have

$$\|g\|_K \leq M\|g\|_{A|K}$$

and so $A|K$ is continuously injected into $C(K)$.

Clearly we may regard the Stone-Čech compactification of $\mathbb{N} \times K$ as a closed subset of $\overline{\mathbb{N} \times X}$. To show that $A|K$ is ultraseparating, i.e. that $(A|K)^{\sim}$ separates the points of $\overline{\mathbb{N} \times K}$, it is then enough to show that $\tilde{A}|\overline{\mathbb{N} \times K} \subset (A|K)^{\sim}$, because $\tilde{A}$ separates the points of $\overline{\mathbb{N} \times X}$ and so surely the points of the subset $\overline{\mathbb{N} \times K}$. But this amounts to showing that if $f_n \in A$ and $\sup_n \|f_n\|_A < \infty$, then $\sup_n \|f_n|K\|_{A|K} < \infty$ (and this is obvious), for if f is the extension of $\{f_n\}$ to a continuous function on $\overline{\mathbb{N} \times X}$ then evidently $f|\overline{\mathbb{N} \times K}$ is the extension of $\{f_n|K\}$.

<u>Lemma 4.11</u>

Let X be a compact Hausdorff space, A a uniformly closed subalgebra of $C(X)$ which separates the points of X and contains the constants. If $\operatorname{Re} A$ is uniformly dense in $C_{\mathbb{R}}(X)$, then A is ultraseparating.

<u>Proof</u>: If $f \in C_{\mathbb{R}}(X)$ and $f > 0$ on X, then $u = \log\circ f \in C_{\mathbb{R}}(X)$. Given $\epsilon > 0$ there is then a $g \in A$ with

$$\|u - \operatorname{Re} g\|_X < \epsilon.$$

Then a simple application of the Mean Value Theorem (to $e^{u(x)} - e^{\operatorname{Re} g(x)}$ for each $x \in X$) gives

$$\|e^u - e^{\operatorname{Re} g}\|_X \le \|u - \operatorname{Re} g\|_X \cdot e^{\|u - \operatorname{Re} g\|_X + \|u\|_X}$$

$$\le \epsilon e^{\epsilon}\|f\|_X,$$

that is,

(1) $$\|f - |e^g|\|_X \le \epsilon e^{\epsilon}\|f\|_X.$$

If $x,y \in \overline{\mathbb{N} \times X}$ and $x \ne y$, Urysohn provides a $\varphi \in C(\overline{\mathbb{N} \times X})$ with $\varphi(x) = 4$, $\varphi(y) = 1$ and $\varphi > 0$. Regard φ as $\{\varphi_n\} \in \ell_\infty(\mathbb{N}, C(X))$. Thus each $\varphi_n > 0$ and we may apply (1) to produce $g_n \in A$ such that the element $h_n = e^{g_n}$ of A satisfies

(2) $$\|\varphi_n - |h_n|\|_X \le 1.$$

It follows that $\{h_n\}$ is bounded (since $\{\varphi_n\}$ is) and so belongs to $\ell_\infty(\mathbb{N},A)$. Let h be its continuous extension to $\overline{\mathbb{N} \times X}$. By (2) and the densensss of $\mathbb{N} \times X$ in $\mathbb{N} \times X$ we have

$$\|\varphi - |h|\|_{\overline{\mathbb{N} \times X}} \le 1$$

and so

$$|4 - |h(x)|| = |\varphi(x) - |h(x)|| \le 1$$

$$|1 - |h(y)|| = |\varphi(y) - |h(y)|| \le 1.$$

It follows $|h(x)| \ne |h(y)|$. As $h \in \ell_\infty(\mathbb{N},A) = \tilde{A}$, we see that $\tilde{A}$ separates the points of $\overline{\mathbb{N} \times X}$.

Lemma 4.12

(Bernard [5]) Let X be a compact Hausdorff space, A and B Banach algebras lying in $C(X)$. If A is ultraseparating, B is conjugate closed and $1 \in A \subset B$, then $B = C(X)$.

Proof: As noted before (III, p. 1) we have $\| \ \|_X \le \| \ \|_A$ and $\| \ \|_X \le \| \ \|_B$, so it is easy to apply the Closed Graph Theorem to see that the inclusion of A into B is continuous. Therefore $\tilde{A} \subset \tilde{B}$. Since $\tilde{A}$ separates the points of $\overline{\mathbb{N} \times X}$, so does $\tilde{B}$. Moreover an application of the (real) Closed Graph Theorem in B, using as before $\| \ \|_X \le \| \ \|_B$, shows that conjugation is continuous in B. This clearly implies $\tilde{B}$ is conjugate closed: if $\|\bar{f}\|_B \le M\|f\|_B$ for all $f \in B$, then $\{f_n\} \in \ell_\infty(\mathbb{N},B)$ implies $\{\bar{f}_n\} \in \ell_\infty(\mathbb{N},B)$ with $\sup_n\|\bar{f}_n\|_B \le M \sup_n\|f_n\|_B$, i.e. with $\|\{\bar{f}_n\}\|_{\tilde{B}} \le M\|\{f_n\}\|_{\tilde{B}}$. Thus $\tilde{B}$ is (naturally identified with) a conjugate closed point-separating subalgebra of $C(\overline{\mathbb{N} \times X})$ which contains the constants. By Stone-Weierstrass $\tilde{B}$ is dense in $C(\overline{\mathbb{N} \times X}) = \ell_\infty(\mathbb{N},C(X)) = \tilde{C}(X)$ and therefore $B = C(X)$ by Bernard's lemma (4.5).

Theorem 4.13

(Bernard [5]) Let X be a compact Hausdorff space, A a Banach algebra lying in $C(X)$ which contains the constants and is ultraseparating. If $\mathrm{Re}\, A$ is closed under multiplication, then $A = C(X)$.

Proof: Let $B = \mathrm{Re}\, A + i\mathrm{Re}\, A$. Since $\mathrm{Re}\, A$ is closed under multiplication, B is a subalgebra of $C(X)$. Let $E = \{f \in A\colon \mathrm{Re}\, f = 0\}$. Then E is a real linear space and is closed in A because $\|\text{Re } f\|_\infty \le \|f\|_\infty \le \|f\|_A$ (III, p. 1). Therefore

$$\|u\| = \inf\{\|u + iv\|_A : u + iv \in A\} \qquad u \in \text{Re } A$$

is a quotient norm in $\text{Re } A$, hence is complete. Moreover $\|\ \|_X \le \|\ \|$. (Cf. IV(c), p. 3) The reader can therefore easily mimic the proof of lemma 3.3 (i.e. a Closed Graph argument) to see that multiplication in $\text{Re } A$ is jointly $\|\ \|$-continuous. Then $\|u + iv\| = \|u\| + \|v\|$ defines a complete real linear space norm in B and $\|f\|_B = \sup_\theta \|e^{i\theta} f\|$ is an equivalent complex linear space norm. Again an imitation of the proof of lemma 3.3 shows that multiplication in B is jointly $\|\ \|_B$-continuous. It follows that B is a Banach algebra. Therefore all the hypotheses of the last lemma are met and it follows that $B = C(X)$. In particular $\text{Re } A = \text{Re } B = \text{Re } C(X) = C_{\mathbb{R}}(X)$. As A is ultraseparating, it certainly separates the points of X, so Bernard's Theorem (4.6) applies and gives $A = C(X)$.

Corollary 4.14

(Wermer) Let X be a compact Hausdorff space, A a uniformly closed point-separating subalgebra of $C(X)$ which contains the constants. If $\text{Re } A$ is closed under multiplication, then $A = C(X)$.

Proof: By the real Stone-Weierstrass Theorem $\text{Re } A$ is uniformly dense in $C_{\mathbb{R}}(X)$. Therefore A is ultraseparating by lemma 4.11 and the conclusion follows from the last theorem.

Definition 4.15

Let X be a topological space, $A \subset C(X)$, $S \subset \mathbb{R}$ and $\varphi: S \to \mathbb{R}$ a function.

(i) Say φ operates in A if $\varphi \circ f \in A$ whenever $f \in A$ and $f(X) \subset S$.

(ii) If A is a normed linear space say φ operates boundedly in A if for every $\varepsilon > 0$ there is an $M(\varepsilon) > 0$ such that $\varphi \circ f \in A$ and $\|\varphi \circ f\|_A \leq M(\varepsilon)$ whenever $f \in A$, $f(X) \subset S$, and $\|f\|_A \leq \varepsilon$.

Lemma 4.16

Let X be a compact Hausdorff space, E a uniformly closed vector subspace of $C_{\mathbb{R}}(X)$ which contains the constants. Let $\varphi, h: \mathbb{R} \to \mathbb{R}$ be continuous with φ compactly supported. If h operates in E then so does $\varphi * h$.

Proof: It is a routine exercise in epsilonics to move from the integral definition of $\varphi * h$ to uniform approximations thereto on compact subsets of $\mathbb{R}$ by linear combinations of translates of h. Since E contains the constants and h operates in E, any translate of h also operates. So for a given $f \in E$, $(\varphi * h) \circ f$ is uniformly approximable on X by linear combinations of $h \circ (f - t)$ for various $t \in \mathbb{R}$, hence $(\varphi * h) \circ f \in E$.

A less direct proof (Hahn-Banach, Riesz Representation and Fubinito all intervene) but one of a type which occurs frequently elsewhere in this monograph (and so would be pointless to eschew here) is as follows. It suffices to show that $\int_X (\varphi * h) \circ f d\mu = 0$ for every $f \in E$ and every finite regular Borel measure μ on X which annihilates E. For any $t \in \mathbb{R}$ we have $h\circ(f - t) \in E$, as noted above, and so for such a μ

$$(*) \qquad \int_X h\circ(f - t)d\mu = 0 \qquad \forall t \in \mathbb{R}.$$

The most elemental form of Fubini's Theorem (finite regular measures, continuous functions) gives

$$\int_X (\varphi * h)\circ f d\mu = \int_X (\varphi * h)(f(x))d\mu(x) = \int_X [\int_{\mathbb{R}} \varphi(t)h(f(x)-t)dt]d\mu(x)$$

$$= \int_{\mathbb{R}} \varphi(t)[\int_X h(f(x)-t)d\mu(x)]dt$$

$$= \int_{\mathbb{R}} \varphi(t)[\int_X h\ (f-t)d\mu]dt$$

$$= 0 \quad \text{by } (*).$$

<u>Lemma 4.17</u>

Let X be a compact Hausdorff space, E a uniformly closed subset of $C_{\mathbb{R}}(X)$. If $\varphi_n : \mathbb{R} \to \mathbb{R}$ operate in E and $\varphi_n \to \varphi$ uniformly on compact subsets of $\mathbb{R}$, then φ operates in E.

<u>Proof</u>: Let $f \in E$. $f(X)$ is a compact subset of $\mathbb{R}$ so $\varphi_n \to \varphi$ uniformly on $f(X)$, i.e. $\varphi_n\circ f \to \varphi\circ f$ uniformly on X. We have $\varphi_n\circ f \in E$ by assumption on φ_n and E is uniformly closed. So $\varphi\circ f \in E$.

<u>Lemma 4.18</u>

If $a < b$ are real numbers, there is a C^∞ function which is supported in (a,b) and strictly positive there.

<u>Proof</u>: For $c \in \mathbb{R}$ define

$$(1) \qquad \varphi_c(t) = \begin{cases} e^{\frac{1}{t-c}} & t < c \\ 0 & t \geq c. \end{cases}$$

The reader may confirm by induction that for each non-negative integer n there is a polynomial P_n such that

$$(2) \qquad \varphi_c^{(n)}(t) = P_n(\tfrac{1}{t-c})\varphi_c(t) \qquad \forall t < c.$$

Since $\lim_{x\to\infty} x^m e^{-x} = 0$ for every integer m, it follows from (2) that

$$(3) \qquad \lim_{t \nearrow c} \varphi_c^{(n)}(t) = 0 \qquad n = 0,1,2,\ldots$$

Therefore $\varphi_c \in C^\infty(\mathbb{R})$. Note that φ_c is supported in $(-\infty,c)$ and is strictly positive there. Therefore the function φ defined by

$$\varphi(t) = \varphi_b(t)\varphi_{-a}(-t) \qquad t \in \mathbb{R}$$

has the desired properties.

<u>Lemma 4.19</u>

Let $h,\varphi\colon \mathbb{R} \to \mathbb{R}$ be continuous, φ compactly supported and continuously differentiable. Then $\varphi * h$ is differentiable and $(\varphi*h)' = \varphi' * h$.

<u>Proof</u>: Let φ be supported in $[a,b]$, let $x_0 \in \mathbb{R}$ and $\epsilon > 0$ be given. The continuous function h is bounded on the compact set $[x_0-b-1, x_0-a+1]$, say

$$(1) \qquad |h(t)| < M \qquad \forall t \in [x_0-b-1, x_0-a+1].$$

The function φ' is uniformly continuous on $\mathbb{R}$ (being supported in the compact set $[a,b]$) and so there exists $0 < \delta(\epsilon) < 1$ such that

$$(2) \qquad |u-v| < \delta(\epsilon) \Rightarrow |\varphi'(u)-\varphi'(v)| < \frac{\epsilon}{M(b-a+2)}.$$

If we consider only x with $|x-x_0| < 1$, then

$$\frac{(\varphi*h)(x)-(\varphi*h)(x_0)}{x-x_0} - (\varphi'*h)(x_0) = \int_{-\infty}^{\infty}\left[\frac{\varphi(x-t)-\varphi(x_0-t)}{x-x_0} - \varphi'(x_0-t)\right]h(t)dt$$

$$(3) \qquad = \int_{-\infty}^{\infty}[\varphi'(x_t-t)-\varphi'(x_0-t)]h(t)dt$$

for some x_t between x and x_0, by the Mean Value Theorem. Since φ' is supported in $[a,b]$ and $|x_t-x_0| \le |x-x_0| < 1$, we get from (3)

$$\left|\frac{(\varphi*h)(x)-(\varphi*h)(x_0)}{x-x_0} - (\varphi'*h)(x_0)\right| = \left|\int_{x_0-b-1}^{x_0-a+1}[\varphi'(x_t-t)-\varphi'(x_0-t)]h(t)dt\right|$$

$$(4) \qquad \overset{(1)}{\le} M\int_{x_0-b-1}^{x_0-a+1}|\varphi'(x_t-t)-\varphi'(x_0-t)|dt.$$

But

$$|(x_t-t) - (x-t)| = |x_t-x| \le |x_0-x| \qquad \forall t \in \mathbb{R}.$$

Therefore if $|x-x_0| < \delta(\epsilon)$ it follows from (2) that

$$|\varphi'(x_t-t) - \varphi'(x_0-t)| < \frac{\epsilon}{M(b-a+2)} \qquad \forall t \in \mathbb{R}$$

and so (4) gives

$$\left|\frac{(\varphi*h)(x)-(\varphi*h)(x_0)}{x-x_0} - (\varphi'*h)(x_0)\right| \le M\int_{x_0-b-1}^{x_0-a+1}\frac{\epsilon}{M(b-a+2)} = \epsilon,$$

holding whenever $|x-x_0| < \delta(\epsilon)$.

Lemma 4.20

Let $\varphi, h: \mathbb{R} \to \mathbb{R}$ be continuous, with φ non-negative, supported in $[-1,1]$ and $\int_{-1}^{1}\varphi = 1$. For each positive integer n, set $\varphi_n(t) = n\varphi(nt)$. Then $\varphi_n*h \to h$ uniformly on compact subsets of $\mathbb{R}$.

Proof: Note that

(1) φ_n is supported in $[-\frac{1}{n},\frac{1}{n}]$

and

(2) $\int_{-\infty}^{\infty}\varphi_n = 1 \quad \forall n.$

Therefore

$$(\varphi_n * h)(x) - h(x) = \int_{-\infty}^{\infty} h(x-t)\varphi_n(t)dt - h(x)\int_{-\infty}^{\infty}\varphi_n(t)dt$$

$$= \int_{-\infty}^{\infty}[h(x-t) - h(x)]\varphi_n(t)dt$$

(3) $$= \int_{-\frac{1}{n}}^{\frac{1}{n}} [h(x-t) - h(x)]\varphi_n(t)dt.$$

Given $\epsilon > 0$ and $M > 0$ the uniform continuity of h on $[-M-1,M+1]$ implies that there is a positive integer $n(M,\epsilon)$ such that

(4) $|h(x-t)-h(x)| < \epsilon \quad \forall x \in [-M,M] \quad \forall |t| < 1/n(M,\epsilon).$

It follows from (4) and (3) that for $n > n(M,\epsilon)$ and all $x \in [-M,M]$

$$|(\varphi_n * h)(x)-h(x)| \le \int_{-\frac{1}{n}}^{\frac{1}{n}} |h(x-t)-h(x)|\varphi_n(t)dt$$

$$\le \int_{-\frac{1}{n}}^{\frac{1}{n}} \epsilon\varphi_n(t)dt = \epsilon \quad \text{by (1) and (2).}$$

Theorem 4.21

(deLeeuw & Katznelson [31]) Let X be a compact Hausdorff space, E a uniformly closed point-separating vector subspace of $C_{\mathbb{R}}(X)$ which contains the constants. If any continuous non-affine function on $\mathbb{R}$ operates in E, then $E = C_{\mathbb{R}}(X)$.

<u>Proof</u>: It suffices to show that E contains the square of each of its elements. For then E is an algebra: $fg = \frac{(f+g)^2 - f^2 - g^2}{2}$ and we may cite real Stone-Weierstrass.

Let $h: \mathbb{R} \to \mathbb{R}$ operate in E and be continuous and non-affine, i.e. for some $a,b \in \mathbb{R}$ and some $0 < \lambda < 1$

(1) $$h(\lambda a + (1-\lambda)b) \neq \lambda h(a) + (1-\lambda)h(b).$$

Use lemmas 4.18 and 4.20 to produce a compactly supported C^∞ function φ with $|\varphi * h - h|$ so small at a,b and $\lambda a + (1-\lambda)b$ that the inequality (1) holds as well for the function $H = \varphi * h$, i.e.

(2) $$H(\lambda a + (1-\lambda)b) \neq \lambda H(a) + (1-\lambda)H(b).$$

By lemma 4.19 the function H is C^∞ and by (2) is not affine. Therefore H'' is not identically 0. Pick $t_0 \in \mathbb{R}$ with $H''(t_0) \neq 0$ and consider

(3) $$\Phi(t) = \frac{2}{H''(t_0)}[H(t + t_0) - H'(t_0)t - H(t_0)] \qquad t \in \mathbb{R}.$$

Then Φ is C^∞ but not affine. Moreover

(4) $$\Phi(0) = \Phi'(0) = 0 \quad \text{and} \quad \Phi''(0) = 2.$$

By lemma 4.16 the function $H = \varphi * h$ operates in E because h does and so (since $1 \in E$) the function Φ operates in E too. Taylor's Theorem provides for each $t \in \mathbb{R}$ a $\theta(t)$ with

(5) $$|\theta(t)| \leq |t|$$

such that

$$\Phi(t) = \Phi(0) + \Phi'(0)t + \frac{\Phi''(0)}{2!}t^2 + \frac{\Phi'''(\theta(t))}{3!}t^3 \qquad t \in \mathbb{R}$$

(6) $$\overset{(4)}{=} t^2 + t^2\varepsilon(t) \qquad t \in \mathbb{R}$$

where $\varepsilon(t) = \frac{1}{6}t(\Phi''' \circ \theta)(t)$. Notice that by (5) θ is continuous at 0 so ϵ is also

(7) $$\lim_{t\to 0} \epsilon(t) = 0.$$

If for each positive integer n we set

$$\Phi_n(t) = n^2\Phi(\tfrac{t}{n}),\ \epsilon_n(t) = \epsilon(\tfrac{t}{n}) \qquad t \in \mathbb{R}$$

then Φ_n operates in E and from (6)

(8) $$\Phi_n(t) = t^2 + t^2\epsilon_n(t).$$

Notice that by (7) we have

$$\lim_{n\to\infty} \epsilon_n = 0 \text{ uniformly on compact subsets of } \mathbb{R}.$$

In particular if $f \in E$ then $\epsilon_n \circ f \to 0$ uniformly on X, and so

$$\|f^2 - \Phi_n \circ f\|_X \overset{(8)}{=} \|f^2 \epsilon_n \circ f\|_X \le \|f\|_X^2 \|\epsilon_n \circ f\|_X \to 0.$$

Since $\Phi_n \circ f \in E$ and E is closed, it follows that $f^2 \in E$.

<u>Lemma 4.22</u>

Let X be a compact Hausdorff space, $E \subset C_{\mathbb{R}}(X)$ a Banach space continuously injected in $C_{\mathbb{R}}(X)$ which contains the constants, separates points and is a lattice. Then each $x \in X$ with at most finitely many exceptions has a compact neighborhood K_x such that $\varphi(t) = |t|$ operates boundedly in $E|K_x$ with its quotient norm.

<u>Proof:</u> Since $|f| = f\vee 0 - f\wedge 0$ we see that φ operates in E. Let N denote the norm in E and for each compact $F \subset X$ let N_F denote the quotient norm induced in $E|F$ by N:

(1) $$N_F(f) = \inf\{\|g\|_N : g \in E, g|F = f\} \qquad \forall f \in E|F.$$

Notice first that if K compact $\subset U$ open $\subset X$, then there is in E a function u with $u(X) \subset [0,1]$, $u(K) = 1$, $u(X\backslash U) = 0$. Indeed $1 \in E$ and E point separating means that for each $x \in K$, $y \in X\backslash U$ there is a function $f_{x,y} \in E$ with $f_{x,y}(x) > 1$, $f_{x,y}(y) < 0$. Thus

(2) $\quad x \in f_{x,y}^{-1}(1,\infty)$ open, $y \in f_{x,y}^{-1}(-\infty,0)$ open.

Cover compact K with, say, $f_{x_1,y}^{-1}(1,\infty),\ldots,f_{x_n,y}^{-1}(1,\infty)$ and consider

$$f_y = f_{x_1,y} \vee \ldots \vee f_{x_n,y} \in E.$$

This function satisfies

$$K \subset f_y^{-1}(1,\infty),\; y \in f_y^{-1}(-\infty,0) \text{ open.}$$

Cover the compact set $X\backslash U$ with, say, $f_{y_1}^{-1}(-\infty,0),\ldots,f_{y_m}^{-1}(-\infty,0)$ and set

$$f = f_{y_1} \wedge \ldots \wedge f_{y_m} \in E.$$

This function satisfies

$$K \subset f^{-1}(1,\infty), X\backslash U \subset f^{-1}(-\infty,0).$$

Finally let

$$u = (f \wedge 1) \vee 0 \in E.$$

This function satisfies

$$K \subset u^{-1}\{1\},\; X\backslash U \subset u^{-1}\{0\},\; u(X) \subset [0,1].$$

It follows that any $g \in E|K$ has an extension $\tilde{g} \in E$ which vanishes on $X\backslash U$. By considering separately $g \vee 0$ and $-(g \wedge 0)$ and scaling, it suffices to demonstrate this for g satisfying $g(K) \subset [0,1]$. If then f is any extension and u is as above, the function $\tilde{g} = (f \vee 0) \wedge u$

has the desired properties.

We now use the hypothesis that E is continuously injected into $C_{\mathbb{R}}(X)$, that is, the supremum norm is dominated by a multiple of the given norm in E. This evidently makes evaluations at the points of X continuous linear functionals on E and so the set $k(X\backslash U)$ of $f \in E$ which vanish on $X\backslash U$ is a closed vector subspace of E. As noted in the last paragraph, the map $g \to g|K$ maps $k(X\backslash U)$ <u>onto</u> $E|K$, and of course continuously. Therefore by the Open Map Theorem, there is a constant $M(K,U)$ such that

(3) $\forall g \in E|K \ \exists \tilde{g} \in k(X\backslash U)$ with $\tilde{g}|K = g$ & $N(\tilde{g}) \leq M(K,U)N_K(g)$.

We now prove the assertion of the lemma by <u>reductio ad absurdum</u>. Suppose X_0 is an infinite subset of X such that φ operates boundedly in no neighborhood of any point of X_0. Let x be a limit point of X_0, x_1 any point of $X_0\backslash\{x\}$, U_1, V_1 disjoint open neighborhoods of x_1 and x respectively. If $x_1,\ldots,x_n \in X_0$ and disjoint open neighborhoods $U_1,\ldots,U_n,V_n$ of $x_1,\ldots,x_n,x$ have been selected, let $x_{n+1} \in V_n \cap (X\backslash\{x\})$ and U_{n+1},V_{n+1} be disjoint open neighborhoods of x_{n+1},x respectively in V_n. We have inductively constructed

(4) $x_1,x_2,\ldots \in X_0$, disjoint open neighborhoods U_n of x_n.

Let $K_n \subset U_n$ be a compact neighborhood of x_n. Since φ does not operate boundedly in $E|K_n$ (but does operate in E) there exists $g_n \in E|K_n$ such that

(5) $$N_{K_n}(g_n) \leq \frac{1}{2^n M(K_n,U_n)}$$

and

(6) $\qquad N_{K_n}(|g_n|) \geq n.$

According to (3) then there is $\tilde{g}_n \in k(X\backslash U_n)$, that is with

(7) $\qquad \tilde{g}_n(X\backslash U_n) = 0,$

such that

(8) $\qquad \tilde{g}_n|K_n = g_n$

and

(9) $\qquad N(\tilde{g}_n) \leq M(K_n,U_n)N_{K_n}(g_n).$

It follows from (9) and (5) that $N(\tilde{g}_n) \leq 2^{-n}$ and therefore $\tilde{g} = \sum_{n=1}^{\infty} \tilde{g}_n$ converges in E. As E is a lattice, $|\tilde{g}| \in E$.

Moreover

(10) $\qquad |\tilde{g}||K_n = |\tilde{g}_n||K_n \overset{(8)}{=} |g_n| \qquad n = 1,2,\ldots,$

because if $k \neq n$ we have $K_n \subset U_n \subset X\backslash U_k$ whence $\tilde{g}_k(K_n) = 0$ by (7). From (10) and the definition of the quotient norm we have

$$N_{K_n}(|g_n|) \leq N(|\tilde{g}|) \qquad n = 1,2,\ldots.$$

The validity of this inequality for $n > N(|\tilde{g}|)$ is in contradiction with (6).

<u>Lemma 4.23</u>

Let X be a compact Hausdorff space, $E \subset C_{\mathbb{R}}(X)$ a real Banach space continuously injected into $C_{\mathbb{R}}(X)$, which contains the constants and is ultraseparating. If any continuous non-affine function on $\mathbb{R}$ operates boundedly in E, then $E = C_{\mathbb{R}}(X)$.

<u>Proof:</u> Let φ be a continuous function on $\mathbb{R}$ which operates boundedly in E. That is, for each $\epsilon \geq 0$ there is

a $M(\epsilon) > 0$ such that

(1) $u \in E$ & $\|u\|_E \le \epsilon \Rightarrow \varphi \circ u \in E$ & $\|\varphi \circ u\|_E \le M(\epsilon)$.

If $\{u_n\} \in \tilde{E} = \ell_\infty(\mathbb{N}, E)$, it follows at once from (1) that $\{\varphi \circ u_n\} \in \tilde{E}$ too--and indeed $\|\{\varphi \circ u_n\}\|_{\tilde{E}} \le M(\|\{u_n\}\|_{\tilde{E}})$. Thus

(2) φ operates in $\tilde{E}$.

Consider $\tilde{E}$ as a subspace of $C_{\mathbb{R}}(\overline{\mathbb{N} \times X})$, $\overline{\mathbb{N} \times X}$ the Stone-Čech compactification of $\mathbb{N} \times X$. If bar denotes uniform closure, then it follows from (2) that

(3) φ operates in $\bar{\tilde{E}}$.

Indeed if $f_k \in \tilde{E}$ and converge to f_0 uniformly on $\overline{\mathbb{N} \times X}$, then there is a compact subset K of $\mathbb{R}$ such that $f_k(\overline{\mathbb{N} \times X}) \subset K$ for all $k = 0,1,2,\ldots$ Since φ is uniformly continuous on the compact set K, it follows from the fact that f_k converges to f_0 uniformly on $\overline{\mathbb{N} \times X}$ that $\varphi \circ f_k$ converges to $\varphi \circ f_0$ uniformly on $\overline{\mathbb{N} \times X}$. As each $\varphi \circ f_k \in \tilde{E}$ by (2) we see that $\varphi \circ f_0 \in \bar{\tilde{E}}$.

The ultraseparation hypothesis means that $\tilde{E}$, and so also $\bar{\tilde{E}}$, separates the points of $\overline{\mathbb{N} \times X}$. If φ is not affine it follows from (3) and the Katznelson-deLeeuw Theorem (4.21) that $\bar{\tilde{E}} = C_{\mathbb{R}}(\overline{\mathbb{N} \times X})$. Hence $\tilde{E}$ is dense in $C_{\mathbb{R}}(\overline{\mathbb{N} \times X}) = \ell_\infty(\mathbb{N}, C_{\mathbb{R}}(X)) = \tilde{C}_{\mathbb{R}}(X)$. It follows from Bernard's lemma (4.5) that $E = C_{\mathbb{R}}(X)$.

Corollary 4.24

(Bernard [6]) Let X be a compact Hausdorff space, A a uniformly closed point-separating subalgebra of $C(X)$ which contains the constants. If some non-affine continuous function operates boundedly in Re A with its quotient norm, then $A = C(X)$.

<u>Proof</u>: Let $\varphi: \mathbb{R} \to \mathbb{R}$ be continuous, non-affine and operate boundedly in Re A with its quotient norm. Then, as shown in the course of the last proof, φ operates in the uniform closure $\overline{\text{Re } A}$ of Re A. Since Re A separates points it follows from the Katznelson-deLeeuw Theorem (4.21) that $\overline{\text{Re } A} = C_{\mathbb{R}}(X)$. Then by lemma 4.11, A is ultraseparating.

We next deduce from this last fact that Re A in its quotient norm is also ultraseparating. Notice first that this notion makes sense because for any $f \in A$

$$(*) \quad \|\text{Re } f\|_X \leq \inf\{\|g\|_X : g \in A, \text{ Re } g = \text{Re } f\} = \|\text{Re } f\|_{\text{Re } A}$$

(by definition of the quotient norm in Re A) and therefore Re A is continuously injected into $C_{\mathbb{R}}(X)$ as required in definition 4.9. Now let $x, y \in \overline{\mathbb{N} \times X}$, $x \neq y$ be given. Since A is ultraseparating there is $f \in \tilde{A}$ with $f(x) \neq f(y)$. $\tilde{A}$ being a complex linear space, we may suppose without loss of generality that $\text{Re } f(x) \neq \text{Re } f(y)$. Now f is the extension to $\overline{\mathbb{N} \times X}$ of some $\{f_n\} \in \ell_\infty(\mathbb{N}, A)$; say, $\sup_n \|f_n\|_A \leq M$.

Since the norm in Re A is the quotient norm, we have $\|\text{Re } f_n\|_{\text{Re } A} \overset{(*)}{\leq} \|f_n\|_A$. Therefore $\{\text{Re } f_n\} \in \ell_\infty(\mathbb{N}, \text{Re } A) \subset \ell_\infty(\mathbb{N}, C_{\mathbb{R}}(X))$. If h is the continuous extension to $\overline{\mathbb{N} \times X}$ of $\{\text{Re } f_n\}$, then $h = \text{Re } f$ on the dense subset $\mathbb{N} \times X$ of $\overline{\mathbb{N} \times X}$ and so $h(x) = \text{Re } f(x) \neq \text{Re } f(y) = h(y)$.

Now that Re A is ultraseparating we can apply the previous lemma to conclude that $\text{Re } A = C_{\mathbb{R}}(X)$. It follows then from any one of several previous results (for example from the Hoffman-Wermer Theorem) that $A = C(X)$.

Theorem 4.25

(Bernard [6]) Let X be a compact Hausdorff space, A a uniformly closed point-separating subalgebra of $C(X)$ which contains the constants. If $\operatorname{Re} A$ is a lattice, then $A = C(X)$.

Proof: If $\operatorname{Re} A$ is a lattice it is uniformly dense in $C_{\mathbb{R}}(X)$. Indeed if μ is a real Borel measure on X which annihilates $\operatorname{Re} A$, K a compact subset of X and $\epsilon > 0$, regularity of μ provides open $U \supset K$ so that $|\mu|(U\backslash K) < \epsilon$. At the beginning of the proof of lemma 4.22 (before the completeness hypothesis on the E there was used) it was shown that there exists $u \in E = \operatorname{Re} A$ with

$$K \subset u^{-1}\{1\},\ X\backslash U \subset u^{-1}\{0\},\ u(X) \subset [0,1].$$

It follows that

$$0 = \int_X u\,d\mu = \int_U u\,d\mu = \mu(K) + \int_{U\backslash K} u\,d\mu$$

$$|\mu(K)| \le \int_{U\backslash K} |u|\,d\mu \le |\mu|(U\backslash K) < \epsilon,$$

holding for all $\epsilon > 0$. Therefore $\mu(K) = 0$, for every compact $K \subset X$. Since μ is regular it follows $\mu = 0$.

As shown in the second paragraph of the proof of Corollary 4.24, it follows from the uniform denseness of $\operatorname{Re} A$ in $C_{\mathbb{R}}(X)$ that $\operatorname{Re} A$ with its quotient norm is ultraseparating. For any compact $K \subset X$, $(\operatorname{Re} A)|K$ is therefore continuously injected in $C_{\mathbb{R}}(K)$ and is ultraseparating by lemma 4.10. Therefore by lemma 4.23 $(\operatorname{Re} A)|K = C_{\mathbb{R}}(K)$ whenever the absolute value operates boundedly in

$E = (\mathrm{Re}\ A)|K$. Consider such a K. We have $(\mathrm{Re}\ A)|K = \mathrm{Re}(A|K)$, a set theoretic banality, and in fact the two norms are also equal: if $u \in \mathrm{Re}(A|K)$ the norm of u in either case is

$$\inf\{\|f\|_X : f \in A,\ \mathrm{Re}\ f|K = u\}.$$

From $\mathrm{Re}(A|K) = C_{\mathbb{R}}(K)$ and Bernard's Theorem 4.6 we get $A|K = C(K)$.

Finally lemma 4.22 ensures that every point $x \in X$ with at most finitely many exceptions has a compact neighborhood K_x such that the absolute value operates boundedly in $(\mathrm{Re}\ A)|K_x$. By the result of the last paragraph $A|K_x = C(K_x)$. It remains only to cite Corollary 2.13 to conclude that $A = C(X)$.

Chapter V

THE THEOREMS OF GORIN AND ČIRKA

This chapter is devoted to some recent work of two Soviet mathematicians. Here topological hypotheses are imposed on the underlying compact Hausdorff space. The proof of the major theorem is quite intricate and the more eclectic reader is reassured that this result is not needed in later chapters. For our first result we begin with a lemma which is elementary but of sufficient intrinsic interest to be dignified with the title "theorem".

Theorem 5.1

If X is a compact Hausdorff space, then $C(X)$ is separable in the uniform norm if and only if X is metrizable.

Proof: ($\Leftarrow$) If d is a metric giving the topology of X, then for each positive integer n, finitely many d-balls of radius $\frac{1}{n}$ cover compact X. If we take the centers of such balls for each n, we have a countable set $\{x_1, x_2, \ldots\}$ which is clearly dense in X. Let f_n be the continuous function on X defined by $f_n(x) = d(x, x_n)$. It is easy to see that the functions f_n separate the points of X and so the algebra which they generate is uniformly dense in $C(X)$ by Stone-Weierstrass. Evidently this algebra is separable. (Dense in it is the countable set of multinomials in the f_n with rational complex coefficients.)

($\Rightarrow$) If $\{f_n : n = 1,2,\ldots\}$ is a countable subset of $C(X)$ which is dense in the uniform norm, then these functions must separate the points of X and so the function $d: X \times X \to \mathbb{R}$ defined by

$$d(x,y) = \sum_{n=1}^{\infty} \frac{1}{2^n} \frac{|f_n(x)-f_n(y)|}{1+\|f_n\|_\infty}$$

is a metric on X. The Hausdorff topology determined by d is weaker than the given compact topology on X and consequently coincides with it.

Lemma 5.2

If A is a commutative Banach algebra with unit 1, then the (multiplicative) group E of exponentials of elements of A is open.

Proof: Letting $B = \{a \in A : \|a-1\| < 1\}$, it suffices to show that $B \subset E$, for then $E = \bigcup_{a \in E} aB$ is open. The series $\sum_{n=1}^{\infty} \frac{-1}{n}(1-a)^n$ converges in A to an element b, say, and by checking the coefficients in the various power series the reader may confirm that $a = e^b$. [For details see Stout [49], p. 7 ff.] In case A is an algebra of functions, as in the application below, this latter tedious procedure can be circumvented as was done in proof of lemma 3.5.

Theorem 5.3

(Gorin [25]) Let X be a compact metric space, A a uniformly closed subalgebra of $C(X)$ which contains 1 such that every strictly positive function in $C_{\mathbb{R}}(X)$ is the modulus of an invertible element of A. Then $A = C(X)$.

Proof: Consider first the case where A is antisymmetric. Let A^{-1} denote the group of invertible elements of A and E the group of exponentials of elements of A. The hypothesis is then that $\log|A^{-1}| = C_{\mathbb{R}}(X)$. If we can show that $E = A^{-1}$, then we shall have

$$\text{Re } A = \log|E| = \log|A^{-1}| = C_{\mathbb{R}}(X)$$

and then $A = C(X)$ follows from any one of several previous results, for example from Corollary 4.3.

To see that $E = A^{-1}$ we shall show that in the contrary case the subgroup E of A^{-1} has uncountable index in A^{-1}. Since by lemma 5.2 E is open, the cosets of E in A^{-1} will constitute uncountably many disjoint open subsets of A. Therefore A and hence $C(X)$ is surely not separable, contradicting Theorem 5.1.

So assume $E \subsetneq A^{-1}$ and fix some $x_0 \in X$ and some $f \in A^{-1} \setminus E$ such that

(1) $\qquad f(x_0) = 1.$

For each $t \in \mathbb{R}$, $t \log|f| \in C_{\mathbb{R}}(X)$ (since f is never 0) which equals $\log|A^{-1}|$, so there is an $f_t \in A^{-1}$ such that $\log|f_t| = t \log|f|$, that is,

(2) $\qquad |f_t| = |f|^t.$

Moreover we may scale f_t to achieve in addition to (2)

(3) $\qquad f_t(x_0) = 1$

and then f_t is uniquely determined. For if $g_1, g_2 \in A^{-1}$, $|g_1| = |g_2|$ and $g_1(x_0) = g_2(x_0)$, then consider $h = g_1/g_2 \in A^{-1}$.

We have $|h| = 1$ so $\overline{h} = 1/h \in A$. By antisymmetry of A then Re h and Im h are each constant. Since $h(x_0) = 1$ these constants are 1 and 0, respectively, so $h = \text{Re } h = 1$. From the uniqueness thus established follows at once that $f_{t+s} = f_t f_s$ and $f_0 = f$ and so $t \to f_t$ is a homomorphism of (the additive group) $\mathbb{R}$ into A^{-1}. Finally if $t \neq s$ then f_t and f_s belong to different cosets of E in A^{-1}. For otherwise $f_{t-s} = f_t f_s^{-1} \in E$, say, $f_{t-s} = e^h$, $h \in A$. Then $e^{h(x_0)} = 1$ and $|f|^{t-s} = |e^h|$, that is, assuming as we thus may that $h(x_0) = 0$, the invertible functions f and $e^{h/t-s}$ agree at x_0 and have identical moduli, hence are identical by the uniqueness result above. But this puts $f \in E$, a contradiction.

It remains to reduce the general situation to the antisymmetric case. But it is obvious that if $\mathcal{K}$ is the maximal A-antisymmetric decomposition of X, then for each $K \in \mathcal{K}$ the algebra $A|K$ (which by part (ii) of Bishop's Theorem (1.2) is uniformly closed in $C(K)$) satisfies all the hypotheses of this theorem. Since $A|K$ is antisymmetric, the result already established shows that $A|K = C(K)$ and then $A = C(X)$ from part (i) of Bishop's Theorem.

Remarks

Gorin shows by an example in [25] that in general the metrizability hypothesis above cannot be dropped. Moreover it is not enough that every positive function in $C_{\mathbb{R}}(X)$ be the modulus of some function in A: In [28] and [39] the authors construct a proper, uniformly closed, point-separating subalgebra A of $C(\overline{\mathbb{N}})$ ($\overline{\mathbb{N}}$ the Stone-Čech compactification of the positive integers $\mathbb{N}$) which contains the constants and is such that each non-negative function in $C_{\mathbb{R}}(\overline{\mathbb{N}})$ is the modulus of some element of A. The proof above is essentially Gorin's, simplified and made more elementary by Sidney and Stout (cf. Stout [49]). Gorin's example may also be found in [49], p. 130.

The next theorem is due to Čirka [15], the (somewhat intimidating) proof below being an elaboration of that in Stout [49].

Theorem 5.4

(Čirka) Let X be a locally connected, compact Hausdorff space, A a point-separating, uniformly closed subalgebra of C(X) which contains the constants. If each function in A is the square of another, then A = C(X).

For the proof we assemble a phalanx of lemmas.

Lemma 5.5

Let Y be a connected topological space, $f_j: Y \to \mathbb{C}$ continuous functions such that $\lim_{j\to\infty} f_j(y_0) = 1$ for some $y_0 \in Y$. If there exist positive integers $n_1 < n_2 < \ldots$ such that $\lim_{j\to\infty} f_j^{n_j} = 1$ uniformly on Y, then $\lim_{j\to\infty} f_j = 1$ uniformly on Y.

Proof: From $f_j^{n_j} \to 1$ uniformly on Y we have $|f_j^{n_j}| > 0$ for all large j, so without loss of generality we can suppose that no f_j is ever zero. Then setting $g_j = f_j/|f_j|$, we have $g_j^{n_j} \to 1$ uniformly on Y. Without loss of generality then

$$|g_j^{n_j} - 1| < 1 \qquad \forall j.$$

Then $-1 \notin g_j^{n_j}(Y)$ and so $g_j(Y) \cap \{e^{2\pi ik/n_j} e^{\pi i/n_j}: k \in Z\} = \emptyset$.

Since $|g_j| = 1$, it follows that

$$g_j(Y) \subset \bigcup_{k=1}^{n_j} e^{\pi i/n_j}\{e^{\pi i\theta}: \frac{k}{n_j} < \theta < \frac{k+1}{n_j}\},$$

a disjoint union of open subsets of the unit circle. Since $g_j(Y)$ is connected, it follows that

$$g_j(Y) \subset e^{\pi i/n_j}\{e^{\pi i\theta}: \frac{k}{n_j} < \theta < \frac{k+1}{n_j}\}$$

for some $k \in Z$. Therefore for all $y \in Y$

$$|g_j(y)-1| \le |g_j(y)-g_j(y_0)|+|g_j(y_0)-1| \le |e^{\pi i/n_j}-1|+|g_j(y_0)-1|.$$

Since $g_j(y_0) = f_j(y_0)/|f_j(y_0)| \to 1$ by hypothesis, we thus have

(1) $\qquad g_j \to 1$ uniformly on Y.

Now given $0 < \epsilon < 1$ there exists j_0 such that

$$|f_j^{n_j} - 1| \le \epsilon \qquad \forall j \ge j_0.$$

Then

$$1 - \epsilon \le |f_j^{n_j}| \le 1 + \epsilon$$

$$1 - \epsilon < (1 - \epsilon)^{1/n_j} \le |f_j| \le (1 + \epsilon)^{1/n_j} < 1 + \epsilon \qquad \forall j \ge j_0,$$

that is

(2) $\qquad |f_j| \to 1$ uniformly on Y.

Since $f_j = g_j|f_j|$, it follows from (1) and (2) that $f_j \to 1$ uniformly on Y.

<u>Lemma 5.6</u>

If Y is a connected topological space, f,g continuous

complex-valued functions on Y which are never zero and satisfy $f(y_0) = g(y_0)$ for some $y_0 \in Y$ and if $f^{2^m} = g^{2^m}$ for some positive integer m, then $f = g$.

<u>Proof</u>: Setting $F = f^{2^{m-1}}$, $G = g^{2^{m-1}}$, we have $F(y_0) = G(y_0)$ and $F^2 = G^2$. Therefore the result follows by induction as soon as the case $m = 1$ is established. Thus suppose $f^2 = g^2$. Then $(f - g)(f + g) = 0$ and so

$$(*) \qquad Y = (f - g)^{-1}\{0\} \cup (f + g)^{-1}\{0\}.$$

Both the sets on the right hand of (*) are closed since $f - g$ and $f + g$ are continuous functions on Y, and they are disjoint because at any common point we would have $f(y) + g(y) = f(y) - g(y)$, whence $g(y) = 0$, contrary to hypothesis. One of these sets must be void, and so the other is all of Y, since otherwise (*) would constitute a disconnection of Y. As $y_0 \in (f - g)^{-1}\{0\}$, it follows that $(f - g)^{-1}\{0\} = Y$.

<u>Lemma 5.7</u>

There exists an $r > 0$ with the following property. For any $a = |a|e^{i\theta} \in \mathbb{C}$ with $|a| \geq 1$, there is in the open disk $D(a,r) = \{z \in \mathbb{C}: |z-a| < r\}$ an analytic logarithm L such that $\operatorname{Im} L$ maps into $(\theta - \frac{\pi}{8}, \theta + \frac{\pi}{8})$.

<u>Proof</u>: Let ℓ be the function defined at the beginning of the proof of lemma 3.4. Recall ℓ is analytic in $D_1 = \{z \in \mathbb{C}: |z| < 1\}$ and satisfies

$$(1) \qquad e^{\ell(z)} = 1 - z \qquad \forall z \in D_1$$

$$(2) \qquad \ell(0) = 0.$$

From (2) and continuity of ℓ there exists $0 < r < 1$ such that

(3) $$\operatorname{Im} \ell(D_r) \subset (-\pi/8, \pi/8)$$

where $D_r = \{z \in \mathbb{C}: |z| < r\}$. From (1) we see

$$ae^{\ell(z)} = a(1 - z) \qquad \forall z \in D_1$$

and therefore, setting

(4) $$L(w) = \log|a| + i\theta + \ell(\tfrac{a-w}{a}) \qquad w \in D(a, |a|),$$

we have

(5) $$e^{L(w)} = w \qquad \forall w \in D(a, |a|)$$

and L is analytic in $D(a, |a|) \supset D(a,r)$. If $w \in D(a,r)$ then $|\frac{a-w}{a}| < \frac{r}{|a|} \le r$ and so (3) and (4) show

(6) $$\operatorname{Im} L(w) = \theta + \operatorname{Im} \ell(\tfrac{a-w}{a}) \in \theta + (-\pi/8, \pi/8).$$

<u>Lemma 5.8</u>

If $t > 0$, then for all sufficiently large positive integers n

$$[(1 + t)^n - 1]^{1/n} \ge 1 + t/2.$$

<u>Proof</u>: We have $0 < 1 - (\frac{1}{1+t})^n < 1$ and so $0 < 1 - (\frac{1}{1+t})^n < [1 - (\frac{1}{1+t})^n]^{1/n}$. It follows that

$$[(1+t)^n-1]^{1/n} = (1+t)[1-(\tfrac{1}{1+t})^n]^{1/n} > (1+t)[1-(\tfrac{1}{1+t})^n] = 1+t-(\tfrac{1}{1+t})^{n-1}.$$

Since $(\frac{1}{1+t})^{n-1} \to 0$ the assertion follows.

<u>Proof of Theorem 5.4</u>

We first show that if $f \in A$, $\epsilon > 0$ and

(1) E compact, connected non-void $\subset |f|^{-1}(1,\infty)$

(2) F compact, connected non-void $\subset |f|^{-1}[0,1)$

then there exists $g \in A$ with

(3) $\|g\|_X < 1$, $\|g\|_F < \epsilon$, $\|1 - g\|_E < 2\epsilon$.

Let $x_0 \in F$ be fixed. For each $m = 1,2,\ldots$ let $f_m \in A$ satisfy

(4) $$f_m^{2^m} = f^{2^m} + 1.$$

Because F is compact and $|f| < 1$ on F, there is a $0 < \theta < 1$ such that $|f| < \theta$ on F. It follows from this and (4) that

$$|f_m^{2^m} - 1| = |f|^{2^m} < \theta^{2^m},$$

(5) $$f_m^{2^m} \to 1 \text{ uniformly on } F \text{ as } m \to \infty.$$

Now if $t \geq 1$ then $t^n - 1 \geq t - 1 \geq 0$ and if $0 \leq t < 1$ then $1 - t^n \geq 1 - t > 0$. It follows that

$$|t-1| \leq |t^n-1| \qquad \forall t \geq 0 \text{ and positive integers } n,$$

and so

$$||f_m| - 1| \leq ||f_m|^{2^m} - 1|.$$

It follows from (5) then that

(6) $$|f_m| \to 1 \text{ uniformly on } F.$$

It is clear from (6) that if we multiply each f_m by the appropriate 2^mth root of unity (so as to rotate the argument of $f_m(x_0)$ into $[0, \frac{2\pi}{2^m})$) then the values at x_0 actually converge to 1. As this does not vitiate (4), we may suppose without loss of generality that in addition to (4) we have

(7) $$f_m(x_0) \to 1.$$

It follows from (5), (7) and lemma 5.5 that

(8) $f_m \to 1$ uniformly on F.

Because $|f| > 1$ on E and E is compact, there is $1 > \delta > 0$ such that

(9) $E \subset S = |f|^{-1}(1 + \delta, \infty)$.

From (4) and the definition of S we have

$$|f_m - 1| \geq |f_m| - 1 = |f^{2^m} + 1|^{1/2m} - 1$$
$$\geq [\,|f|^{2^m} - 1]^{1/2m} - 1$$
$$\geq [(1 + \delta)^{2^m} - 1]^{1/2m} - 1 \quad \text{on } S.$$

It follows from lemma 5.8 that there is an $m_0 \geq 2$ such that

(10) $|f_m - 1| \geq \frac{\delta}{2}$ on S $\quad \forall m \geq m_0$.

Let $r > 0$ be as provided by lemma 5.7. Because X is locally connected and f is continuous, each point of E lies in a connected open neighborhood V such that $\operatorname{diam} f(V) < r$. Of course because of (9) we may require $V \subset S$ as well. Cover compact E with finitely many such neighborhoods:

(11) $E \subset V_1 \cup \ldots \cup V_q \subset S$

(12) V_j open connected, $\operatorname{diam} f(V_j) < r$.

In addition we may require that

(13) $E \cap V_j \neq \emptyset$ for each $j = 1,2,\ldots,q$.

Select $x_j \in V_j$, write $f(x_j) = r_j e^{i\theta_j}$ and let D_j be the open disk of radius r centered at $f(x_j)$. Note that by (12)

(14) $f(V_j) \subset D_j$

and by (9) and (11), $r_j > 1 + \delta$. It follows from lemma 5.7 that there is in D_j a continuous logarithm L_j with

(15) $\quad \operatorname{Im} L_j(D_j) \subset (\theta_j - \frac{\pi}{8}, \theta_j + \frac{\pi}{8})$.

Since L_j is a logarithm we get from (14)

(16) $\quad f(x) = e^{L_j(f(x))} \qquad \forall x \in V_j$.

Now let ℓ be the continuous function defined at the beginning of the proof of lemma 3.4. Recall that

(17) $\quad e^{\ell(z)} = 1 - z \qquad \forall |z| < 1$

(18) $\quad \operatorname{Im} \ell(z) \in (-\pi/2, \pi/2) \qquad \forall |z| < 1.$

Since $|f| > 1$ on S we get from (17)

(19) $\quad 1 + f^{-2^m}(x) = e^{\ell(-f^{-2^m}(x))} \qquad \forall x \in S.$

If we form

(20) $\quad \ell_{j,m}(x) = 2^{-m}\ell(-f^{-2^m}(x)) + L_j(f(x)) \qquad \forall x \in V_j$

then $\ell_{j,m}$ is a continuous complex function on V_j and by (16) and (19) satisfies

(21) $\quad e^{2^m \ell_{j,m}} = [1 + f^{-2^m}]\cdot f^{2^m} = f^{2^m} + 1 \quad \text{in } V_j$

and by (14), (15) and (18) satisfies

(22) $\quad \operatorname{Im} \ell_{j,m} \subset \theta_j + (-\frac{\pi}{8} - \frac{\pi}{2^{m+1}}, \frac{\pi}{8} + \frac{\pi}{2^{m+1}}) \quad \text{in } V_j.$

From (21) and (4)

$$(e^{\ell_{j,m}})^{2^m} = f_m^{2^m} \quad \text{in } V_j.$$

Since V_j is connected it follows from this and lemma 5.6 that

$$f_m = e^{\ell_{j,m} + 2\pi i k 2^{-m}} \qquad \text{in } V_j$$

for some integer $k = k(j,m)$. It follows from this expression for f_m and from (22) that

(23) $\operatorname{Re}[e^{-(2\pi k2^{-m}+\theta_j)i} f_m] > 0$ in V_j, $\forall m \geq 2$.

It follows from (23) that $f_m(V_j)$ is disjoint from $(-\infty,0)$ or from $(0,\infty)$ according as the exponential there has non-negative or negative real part. Moreover by (10), $f_m(S)$ is disjoint from $[-1,1]$ for all $m \geq m_0$. Therefore

(24) f_m-1 is disjoint from either $(-\infty,0]$ or from $[0,\infty)$ on V_j $\forall m \geq m_Q$.

For the next phase of the proof select for each positive integer N and each positive integer m a function $f_{N,m} \in A$ such that

(25) $$f_{N,m}^{2^N} = f_m - 1.$$

We have from (4)

$$\|f_m\|_X^{2^m} \leq \|f\|_X^{2^m} + 1 \leq 2\|f\|_X^{2^m} \leq 2^{2^m}\|f\|_X^{2^m}.$$

Therefore $\|f_m\|_X \leq 2\|f\|_X$ and from (25)

$$\|f_{N,m}\|_X^{2^N} \leq 2\|f\|_X + 1$$

(26) $$\|f_{N,m}\|_X \leq (2\|f\|_X + 1)^{1/2N} \qquad \forall N,m.$$

Also (25) and (10) give

(27) $|f_{N,m}| \geq (\frac{\delta}{2})^{1/2N}$ on S for all N and all $m \geq m_0$.

It follows from (26) and (27) (recall: $0 < \delta < 1$) that

(28) $\lim_{N\to\infty}|f_{N,m}| = 1$ uniformly on S and uniformly in $m \geq m_0$.

Because of (24) and (25), $f_{N,m}(V_j)$ is disjoint from all the rays $[0,\infty)e^{2\pi ik2^{-N}}$ or from all of the rays $[0,\infty)e^{2\pi ik2^{-N}+\pi i2^{-N}}$

$(k \in Z)$ and therefore, as in the proof of lemma 5.5,

$$(29) \qquad \left|\frac{f_{N,m}}{|f_{N,m}|}(x) - \frac{f_{N,m}}{|f_{N,m}|}(x_j)\right| \le |e^{2\pi i 2^{-N}} - 1| \quad \forall x \in V_j, \forall N, \forall m \ge m_0.$$

Apply the diagonal process to produce a sequence $N_1 < N_2 < \dots$ such that for each $j = 1,2,\dots,q$ and all m

$$(30) \qquad \lim_{k\to\infty} f_{N_k,m}(x_j) = \lambda(j,m)$$

exists. It follows from (28) that

$$(31) \qquad |\lambda(j,m)| = 1 \qquad j = 1,2,\dots,q;\ m \ge m_0.$$

As noted before in a similar circumstance, we may, in view of (30) and (31), multiply the $f_{N_k,m}$ by appropriate 2^{N_k}th roots of unity so as to have, without disturbing (25) - (29), $\lim_{k\to\infty} f_{N_k,m}(x_1) = 1$ for every $m \ge m_0$, that is,

$$(32) \qquad \lambda(1,m) = 1 \qquad \forall m \ge m_0.$$

It follows from (28),(29), and (30) that

$$(33) \qquad \lim_{k\to\infty} f_{N_k,m} = \lambda(j,m) \text{ uniformly on } V_j \text{ and uniformly in } m \ge m_0 \text{ for each } j = 1,2,\dots,q.$$

Recall now (11), (12) and (13). Connectedness of E means that V_1 meets one of $V_2,\dots,V_q$. Say V_2. It follows then from (32) and (33) that

$$(34) \qquad \lambda(2,m) = 1.$$

Similarly $V_1 \cup V_2$ meets one of $V_3,\dots,V_q$. Say V_3. It follows then from (32), (34) and (33) that $\lambda(3,m) = 1$. Proceeding in this way we learn finally that

$$\lambda(1,m) = \dots = \lambda(q,m) = 1 \qquad \forall m \ge m_0.$$

It follows from this and (33) that

(35) $\lim_{k\to\infty} f_{N_k,m} = 1$ uniformly on $V_1 \cup \ldots \cup V_q$ and uniformly in $m \geq m_0$.

Finally if $\epsilon > 0$ is given we use (35) [and (11)] to pick k so that

(36) $$\|f_{N_k,m} - 1\|_E < \epsilon \qquad \forall m \geq m_0.$$

Because of (26) we can also require that

(37) $$\|f_{N_k,m}\|_X < 1 + \epsilon \qquad \forall m.$$

Then choose for this k an $m \geq m_0$ according to (8) so that

$$\|f_m - 1\|_F < \epsilon^{2^{N_k}}$$

and it will follow from (25) that

(38) $$\|f_{N_k,m}\|_F < \epsilon.$$

The reader may now confirm that, in view of (37), (38) and (36), the function $g = \frac{f_{N_k,m}}{1+\epsilon}$ does the job promised in (3).

The next phase of the proof is to show that separation like (3) can be achieved for any two disjoint compact subsets E and F of X with no additional hypotheses. So let such a pair be given and $\epsilon > 0$. Since A separates the points of X, for each $x \in E$ and $y \in F$ there is a $f_{x,y} \in A$ with $f_{x,y}(x) \neq f_{x,y}(y)$. By scaling and adding constants ($1 \in A$) we may suppose $0 = |f_{x,y}(x)| < 1 < |f_{x,y}(y)|$. Then there are connected open neighborhoods $U_{x,y}$ of x and $V_{x,y}$ of y such that $|f_{x,y}| < 1$ on $\overline{U}_{x,y}$ and $|f_{x,y}| > 1$ on $\overline{V}_{x,y}$. By a finite covering argument we find open connected sets $U_1,\ldots,U_p$, $V_1,\ldots,V_q$ and functions $f_{ij} \in A$ such that

(39) $$E \subset U_1 \cup \dots \cup U_p, \quad F \subset V_1 \cup \dots \cup V_q$$

(40) $$\overline{U}_i \subset |f_{ij}|^{-1}[0,1), \quad \overline{V}_j \subset |f_{ij}|^{-1}(1,\infty).$$

If $0 < c < 1$ the function

(41) $$\varphi(z) = \frac{z-c}{1-cz} \qquad |z| \le 1$$

is a conformal map of $D_1 = \{z \in \mathbb{C}: |z| < 1\}$ onto itself and satisfies

(42) $$\varphi(1) = 1$$

and $|\varphi(0) + 1| = 1-c$. Therefore we may choose $0 < c < 1$ so that

$$|\varphi(0) + 1| < 2\epsilon/q.$$

Let $0 < \eta < 1$ be so small that

(43) $$|\varphi(z) + 1| < 2\epsilon/q \qquad \text{for all } |z| < \eta$$

and

(44) $$|\varphi(z) - 1| < 2\epsilon \quad \text{for all } z \in D_1 \text{ with } |z-1| < p\eta.$$

Use the result of the first part of the proof to find $g_{ij} \in A$ with

(45) $$\|g_{ij}\|_X < 1, \|g_{ij}\|_{\overline{U}_i} < \eta, \|g_{ij}-1\|_{\overline{V}_j} < \eta \quad i = 1,2,\dots,p; \ j = 1,2,\dots,q.$$

If we set

$$g_j = \prod_{i=1}^{p} g_{ij}$$

then we clearly have

(46) $$\|g_j\|_X < 1$$

and on $\bigcup_{i=1}^{p} U_i$ we have $|g_j| < \eta$, and so by (39)

(47) $$\|g_j\|_E < \eta.$$

On the set V_j we have

$$|g_j-1| \le |\prod_{i=1}^{p} g_{ij} - \prod_{i=2}^{p} g_{ij}| + |\prod_{i=2}^{p} g_{ij} - \prod_{i=3}^{p} g_{ij}| + \ldots + |g_{pj}-1|$$

$$\le |g_{1j}-1| + |g_{2j}-1| + \ldots + |g_{pj} - 1|$$

(recalling that all $\|g_{ij}\|_X < 1$)

$$< p\eta \quad \text{by (45)}$$

(48) $\|g_j - 1\|_{V_j} < p\eta \qquad j = 1,2,\ldots,q.$

Because the power series of φ converges uniformly in the closed disk of radius $\|g_j\|_X < 1$, the function $\varphi \circ g_j \in A$. Form

$$h = \prod_{j=1}^{q} \tfrac{1}{2}(1 - \varphi \circ g_j) \in A.$$

We clearly have

(49) $\|1-\varphi \circ g_j\|_X \le 1 + \|\varphi \circ g_j\|_X < 2 \qquad j = 1,2,\ldots,q$

so

(50) $\|h\|_X < 1.$

If $x \in F$ then $x \in V_j$ for some j and so $|g_j(x) - 1| < p\eta$ by (48). Then by (44), $|1 - \varphi(g_j(x))| < 2\epsilon$. All the other factors in h are of modulus less than 1 by (49), so this gives $|h(x)| < \epsilon$, that is,

(51) $\|h\|_F < \epsilon.$

If $x \in E$ then $|g_j(x)| < \eta$ by (47) so $|\varphi(g_j(x))+1| < 2\epsilon/q$ for each $j = 1,2,\ldots,q$ by (43). Since all the factors in h have modulus less than 1, it follows just as in the derivation of (48) that

$$|h(x)-1| \le \sum_{j=1}^{q} |\tfrac{1}{2}(1 - \varphi \circ g_j)(x) - 1|$$

$$= \sum_{j=1}^{q} \tfrac{1}{2}|\varphi(g_j(x)) + 1| < \sum_{j=1}^{q} \frac{\epsilon}{q} = \epsilon,$$

that is,

(52) $\|h - 1\|_E < \epsilon.$

Finally, with (50) - (52) in hand, the proof that $A = C(X)$ is concluded by an application of Hahn-Banach. We are to show that if μ is a finite regular Borel measure on X which annihilates A, then $\mu = 0$. By regularity it suffices to show $\mu(E) = 0$ for each compact $E \subset X$. Use regularity to choose compact $F_n \subset X \backslash E$ with

(53) $|\mu|(X\backslash F_n \cup E) < 1/n.$

Use (50) - (52) to choose $h_n \in A$ with

(54) $\|h_n\|_X < 1, \ \|h_n\|_{F_n} < 1/n, \ \|h_n - 1\|_E < 1/n.$

It follows from (54) and (53) that h_n converges to the characteristic function χ_E of E almost everywhere with respect to $|\mu|$. By Dominated Convergence and the fact μ annihilates A

$$0 = \int_X h_n d\mu \to \int_X \chi_E d\mu = \mu(E).$$

<u>Corollary 5.9</u>

(Čirka) Let X be a locally connected compact Hausdorff space, A a uniformly closed, point-separating subalgebra of $C(X)$ which contains the constants. If the set $\{g^2 : g \in A\}$ of squares in A is dense in A, then $A = C(X)$.

<u>Proof</u>: To reduce this to the previous result we must show that if $g_n, f \in A$ and $g_n^2 \to f$ uniformly on X, then $f = g^2$ for some $g \in A$. Since the limit of any subsequence of $\{g_n\}$ which happens to converge uniformly on X will

belong to A and have square f, we shall look for such a subsequence. We have $|g_n|^2 \to |f|$ so $(\mathrm{Re}\ g_n)^2 = \frac{1}{2}\mathrm{Re}(|g_n|^2+g_n^2) \to \frac{1}{2}\mathrm{Re}(|f|^2 + f^2) = (\mathrm{Re}\ f)^2$ and $(\mathrm{Im}\ g_n)^2 = \frac{1}{2}\mathrm{Re}(|g_n|^2-g_n^2) \to \frac{1}{2}\mathrm{Re}(|f|^2 - f^2) = (\mathrm{Im}\ f)^2$. Therefore to produce the desired uniformly convergent subsequence of $\{g_n\}$ it suffices to make two <u>successive</u> applications of

<u>Lemma 5.10</u>

Let X be a locally connected compact Hausdorff space $\varphi_n, \psi \in C_{\mathbb{R}}(X)$ and φ_n^2 uniformly convergent to ψ. Then $\{\varphi_n\}$ has a uniformly convergent subsequence.

<u>Proof</u>: (From [49]) For each positive integer K set

$$(1)\qquad E(K) = \psi^{-1}(1/K,\infty).$$

Notice that $\overline{E(K)}$ compact $\subset E(2K)$ open. Therefore by local connectedness and compactness, $\overline{E(K)}$ may be covered by finitely many connected subsets of $E(2K)$. Thus there are only finitely many components of $E(2K)$ which meet $E(K)$. Let them be $C(K,1),\ldots,C(K,q_K)$:

$$(2)\qquad E(K) \subset \bigcup_{j=1}^{q_K} C(K,j) \subset E(2K).$$

For each K and j pick $x_{K,j} \in C(K,j) \cap E(K)$. Use a diagonal argument to produce a subsequence of $\{\varphi_n\}$ which converges at each $x_{K,j}$, say to $\lambda_{K,j}$. Assuming, as we thus may without loss of generality, that $\{\varphi_n\}$ itself converges at each $x_{K,j}$, we will show that $\{\varphi_n\}$ is uniformly Cauchy on X.

Notice that $\lim_{n\to\infty} \varphi_n(x_{K,j}) = \lambda_{K,j}$ and $\varphi_n^2 \to \psi$ imply

$\lambda_{K,j}^2 = \psi(x_{K,j}) > 1/K$, since $x_{K,j} \in E(K)$. Therefore

(3) $|\lambda_{K,j}| > 1/K \qquad j = 1,2,\ldots,q_K;\ K = 1,2,\ldots.$

Given K choose $n(K)$ so large that

(4) $|\varphi_n(x_{K,j}) - \lambda_{K,j}| < \frac{1}{2K} \qquad j = 1,2,\ldots,q_K \qquad \forall n \ge n(K)$

and such that

(5) $\|\varphi_n^2 - \varphi_m^2\|_X < \frac{1}{2K} \qquad \forall n,m \ge n(K)$

and therefore

(6) $\|\varphi_n^2 - \psi\|_X \le \frac{1}{2K} \qquad \forall n \ge n(K).$

We claim that for each $j = 1,2,\ldots,q_K$ and for each $n \ge n(K)$ the function φ_n has throughout the set $C(K,j)$ the same sign as $\lambda_{K,j}$. (Recall that the $\lambda_{K,j}$ are not zero by (3).) For suppose $\lambda_{K,j} > 0$. Then

(7) $\varphi_n(x_{K,j}) \ge \lambda_{K,j} - |\varphi_n(x_{K,j}) - \lambda_{K,j}| \overset{(4)}{\ge} \lambda_{K,j} - \frac{1}{2K} \overset{(3)}{>} \frac{1}{K} - \frac{1}{2K} > 0.$

Since $C(K,j)$ is connected, $\varphi_n(C(K,j))$ is a connected subset of $\mathbb{R}$ which, according to (7), meets $(0,\infty)$. If therefore it also meets $(-\infty,0]$, then it contains 0; that is, $\varphi_n(x) = 0$ for some $x \in C(K,j)$. As $C(K,j) \subset E(2K)$, it follows that $\psi(x) > \frac{1}{2K}$ and so $|\varphi_n^2(x) - \psi(x)| = \psi(x) > \frac{1}{2K}$, in contradiction to (6).

In particular we have, recalling (2),

(8) $\begin{cases} \text{for each } x \in E(K),\ \varphi_n(x) \text{ and } \varphi_m(x) \text{ have the same} \\ \text{sign for all } m,n \ge n(K). \end{cases}$

Consider $x \in X \setminus E(K)$. Then $|\psi(x)| = \psi(x) \le 1/K$ by definition of $E(K)$, so from (6) for $n \ge n(K)$

$$\varphi_n^2(x) \le \psi(x) + \frac{1}{2K} \le \frac{1}{K} + \frac{1}{2K} < \frac{2}{K}.$$

Therefore

(9) $|\varphi_n(x)-\varphi_m(x)| \le |\varphi_n(x)|+|\varphi_m(x)| < 2\sqrt{\frac{2}{K}} \quad \forall n,m \ge n(K), x \in X\backslash E(K).$

If on the other hand, $x \in E(K)$, then for $n,m \ge n(K)$

$$|\varphi_n(x)+\varphi_m(x)| \overset{(8)}{=} |\varphi_n(x)| + |\varphi_m(x)| \overset{(6)}{\ge} 2\sqrt{\psi(x) - \frac{1}{2K}}$$

$$> 2\sqrt{\frac{1}{K} - \frac{1}{2K}} \quad \text{by definition of } x \in E(K)$$

(10) $$= \frac{2}{\sqrt{2K}} \qquad n,m \ge n(K),\ x \in E(K).$$

It follows that

$$|\varphi_n(x) - \varphi_m(x)| = \frac{|\varphi_n^2(x)-\varphi_m^2(x)|}{|\varphi_n(x)+\varphi_m(x)|} \le \frac{\sqrt{2K}}{2}|\varphi_n^2(x) - \varphi_m^2(x)|$$

$$\le \frac{\sqrt{2K}}{2} \cdot \frac{1}{2K} \quad \text{by (5)}$$

(11) $$= \frac{1}{2\sqrt{2K}} \qquad \forall n,m \ge n(K),\ x \in E(K).$$

It follows from (9) and (11) that

$$\|\varphi_n - \varphi_m\|_X \le 2\sqrt{\frac{2}{K}} \qquad \forall n,m \ge n(K).$$

As this is the case for any K, the sequence $\{\varphi_n\}$ is uniformly Cauchy, as desired.

Remarks

As to the status of the local connectedness hypothesis in Theorem 5.4 we remark that some hypothesis beyond the basic compact Hausdorff is needed. For Cole [16] constructs compact metric spaces X and uniformly closed, point-separating proper subalgebras A of $C(X)$ with $1 \in A$ and every function in A the square of another. For the ingenious details the reader may wish to consult the exposition in [49], pp. 194 - 201.

Chapter VI

BOUNDED APPROXIMATE NORMALITY, THE WORK OF BADĒ & CURTIS

Definition 6.1

Let X be a locally compact Hausdorff space, $C_0(X)$ the continuous complex functions vanishing at infinity on X.

(i) If F_1, F_2 are disjoint compact subsets of X and $\epsilon \geq 0$, then a function $f \in C_0(X)$ such that $|f(F_1)| \leq \epsilon$ and $|1-f(F_2)| \leq \epsilon$ is called an ϵ-idempotent with respect to the pair (F_1, F_2).

(ii) Call $E \subset C_0(X)$ regular if E contains a 0-idempotent with respect to each pair $(\{x\}, F)$ with F compact and $x \notin F$.

(iii) Call $E \subset C_0(X)$ ϵ-normal if E contains an ϵ-idempotent with respect to each pair of disjoint compacta in X and call E simply normal if it is 0-normal.

(iv) If E is a normed linear space lying in $C_0(X)$, call E boundedly ϵ-normal if there is a constant K such that the ϵ-idempotents required in (iii) can be found with E-norm less than K.

(v) If E is a normed linear space lying in $C_0(X)$, call E locally boundedly ϵ-normal if X is a union of open sets U for which there exist constants K_U such that E contains an ϵ-idempotent of norm less than K_U for every pair of disjoint compacta in $\overline{U}$.

Under certain conditions regular implies normal (Theorem 6.4 below). We shall show next that for normed algebras, (iv) for some $\epsilon < \frac{1}{2}$ implies (iv) for every $\epsilon > 0$.

Lemma 6.2

Let X be a locally compact Hausdorff space, A a normed algebra lying in $C_0(X)$. If A is boundedly ϵ_0-normal for some $\epsilon_0 < \frac{1}{2}$, then A is boundedly ϵ-normal for every $\epsilon > 0$.

Note

We are asserting only that there is a bound, as per (iv) of the above definition, for each $\epsilon > 0$. We are not asserting that a single bound independent of ϵ exists.

Proof: Given $\epsilon > 0$, lemma 2.3 provides a polynomial q_ϵ such that

$$|q_\epsilon(z) - 1| < \epsilon/2 \qquad \text{if} \quad |z - 1| \le \epsilon_0$$

$$|q_\epsilon(z)| < \epsilon/2 \qquad \text{if} \quad |z| \le \epsilon_0.$$

Then $|q_\epsilon(0)| < \epsilon/2$ and $p_\epsilon = q_\epsilon - q_\epsilon(0)$ is a polynomial without constant term satisfying

(1) $\quad |p_\epsilon(z) - 1| < \epsilon \qquad \text{if} \quad |z - 1| \le \epsilon_0$

(2) $\quad |p_\epsilon(z)| < \epsilon \qquad \text{if} \quad |z| \le \epsilon_0.$

If $p_\epsilon(z) = \sum_{j=1}^{n} c_j z^j$ let $P_\epsilon(z) = \sum_{j=1}^{n} |c_j| z^j$. Let K be a finite constant such that for any two disjoint compact subsets E,F of X, the algebra A contains an ϵ_0-idempotent f for the pair (E,F) with $\|f\|_A \le K$. Then $p_\epsilon \circ f \in A$ (as p has zero constant term) and

(3) $$\|p_\epsilon \circ f\|_A \le p_\epsilon(K).$$

Since $|f(E)| \le \epsilon_0$ and $|1 - f(F)| \le \epsilon_0$ it follows from (1) and (2) that

$$|p_\epsilon \circ f(E)| < \epsilon \quad \text{and} \quad |1 - p_\epsilon \circ f(F)| < \epsilon,$$

that is, $p_\epsilon \circ f$ is an ϵ-idempotent for the pair (E,F).

Remark 6.3

We remark for later use that the version of the last lemma in which all reference to a norm in A is suppressed is also valid: read the above proof with its 9th through 14th lines suppressed.

In connection with normality hypotheses which will appear in several subsequent theorems, we note that normality is implied by certain other natural hypotheses:

Theorem 6.4

Let X be a compact Hausdorff space.

(i) If A is a Banach algebra lying in $C(X)$ which is regular, contains 1 and for which X is the maximal ideal space, then A is normal.

(ii) If A is a Banach algebra lying in $C(X)$ which is normal and conjugate closed, then the separation in the definition of normality can be achieved by functions with range in $[0,1]$.

Proof:

(i) Let E,F be disjoint closed subsets of X and let $I = kF$, $J = kE$ be the ideals of functions in A vanishing on F, E respectively. Note that if $x \in X$ then either $f(x) \neq 0$ for some $f \in I$ or $g(x) \neq 0$ for some $g \in J$. Indeed if $x \in F$ then $x \notin E$ so by regularity there exists $g \in A$ with $g(E) = 0$, $g(x) \neq 0$, i.e. $g \in J$ and $g(x) \neq 0$. Similarly if $x \notin F$. It follows that the ideal $I + J$ generated by $I \cup J$ is contained in no maximal ideal of A, since by hypothesis the maximal ideals of A are all of the form $\{f \in A: f(x) = 0\}$ for some $x \in X$. Therefore $1 \in I + J$. Let, say, $1 = f + g$ where $f \in I$, $g \in J$. Then $f = 1 - g$ vanishes on F and is 1 on E.

(ii) In this case $1 \in A$, because by normality A contains a function which is 0 on $\emptyset$ and 1 on X. If E, F are disjoint closed subsets of X and $f \in A$ is 0 on F and 1 on E, then consider

$$h = \sin^2\circ(\tfrac{\pi}{2}|f|^2) = -\tfrac{1}{4}[\exp\circ(i\,\tfrac{\pi}{2}f\bar{f}) - \exp\circ(-i\,\tfrac{\pi}{2}f\bar{f})]^2$$

which belongs to A (power series). Then h maps X into $[0,1]$, $h(F) = \sin^2(0) = 0$, $h(E) = \sin^2(\frac{\pi}{2}) = 1$.

<u>Theorem 6.5</u>

(Badé & Curtis [3]) Let X, Y be Banach spaces, $k < 1$ and $c > 0$ finite constants, $E \subset Y$ and $\overline{\text{coe}}E$ the closed equilibrated convex hull of E (i.e. $D\cdot\overline{\text{co}}\,E$ where D is the closed unit disk in $\mathbb{C}$). Suppose

(i) $\overline{\text{coe}}\,E \supset B$, the closed unit ball of Y, and let $T: X \to Y$ be a bounded linear operator such that

(ii) for each $y \in E$, there exists $x \in X$ such that

$$\|Tx - y\| \le k, \ \|x\| \le c.$$

Then $TX = Y$. If moreover T is 1-1 then $\|T^{-1}\| \le c(1-k)^{-1}$.

We need the following simple variant of the Separation Theorem.

Lemma 6.6

If Y is a Banach space, $E \subset Y$, B the closed unit ball of Y, then $\overline{\text{coe}}\ E \supset B$ if and only if

$$\|y^*\| \le \sup_{y \in E} |y^*(y)| \qquad \forall y^* \in Y^*.$$

Proof: Note that $\text{coe}\ E$ consists of all finite sums $\sum_{j=1}^{n} c_j y_j$ where $c_j \in \mathbb{C}$, $\sum_{j=1}^{n} |c_j| \le 1$, $y_j \in E$. If $\overline{\text{coe}}\ E \supset B$ then for every $y^* \in Y^*$ and $\epsilon > 0$ we can find $y_0 = \sum_{j=1}^{n} c_j y_j \in \text{coe}\ E$ such that

$$|y^*(y_0)| > \|y^*\| - \epsilon.$$

But then for at least one j we must have

$$|y^*(y_j)| > \|y^*\| - \epsilon$$

(since $\Sigma |c_j| \le 1$) and we see that

$$\sup_{y \in E} |y^*(y)| \ge \|y^*\|.$$

Conversely, suppose $y_0 \in B \backslash \overline{\text{coe}}\ E$. By the Separation Theorem (viz. the Hahn-Banach Theorem) there exists $y_0^* \in Y^*$ such that

$$\text{Re}\ y_0^*(y_0) > \sup\{\text{Re}\ y_0^*(y) : y \in \overline{\text{coe}}\ E\}.$$

Since $y \in \overline{\text{coe}}\ E$ implies $cy \in \overline{\text{coe}}\ E$ for every unimodular complex c, the right side above equals

$$\sup\{|y_0^*(y)|: y \in \overline{\mathrm{coe}}\ E\}$$

and so we get, since $\|y_0\| \leq 1$,

$$\|y_0^*\| \geq \mathrm{Re}\ y_0^*(y_0) > \sup_{y \in E} |y_0^*(y)|.$$

<u>Proof of Theorem 6.5</u>: For any $y^* \in Y^*$ and $y \in E$ select $x \in X$ by (ii) and note that

$$\begin{aligned} |y^*(y)| &\leq |y^*(y - Tx)| + |y^*(Tx)| \\ &\leq \|y^*\|\ \|y - Tx\| + |T^*(y^*)(x)| \qquad (1)\\ &\leq \|y^*\|\cdot k + \|T^*(y^*)\|\ \|x\| \\ &\leq \|y^*\|\cdot k + \|T^*(y^*)\|\cdot c. \end{aligned}$$

Take the supremum over $y \in E$ and get via lemma 6.6

$$\|y^*\| \leq k\|y^*\| + c\|T^*(y^*)\|$$

$$\|T^*(y^*)\| > (1 - k)c^{-1}\|y^*\|. \qquad (2)$$

It follows at once from this inequality that T^* has a closed range. But it is well known and not too difficult that $T^*(Y^*)$ closed implies $T(X)$ closed. (For a brief proof, see Kaufman [35].) However, it follows from (2) that $T(X)$ is dense: if not, there exists $0 \neq y^* \in Y^*$ with $y^*(TX) = T^*(y^*)(X) = 0$, so from (2)

$$0 < \|y^*\| \leq (1 - k)^{-1}c\|T^*(y^*)\| = 0,$$

summarily, $TX = Y$. If now T is 1-1, then T is an onto Banach space isomorphism and so its adjoint T^* is also, and (as shown by simple calculations) $(T^{-1})^* = (T^*)^{-1}$, $\|(T^{-1})^*\| = \|T^{-1}\|$. It follows then from (2) that

$$\|T^{-1}\| = \|(T^{-1})^*\| = \|(T^*)^{-1}\| \overset{(2)}{\leq} (1 - k)^{-1}c.$$

Badé and Curtis [3] make an interesting application of Theorem 6.5 to Helson sets. It depends upon the following fact, of quite independent interest.

Theorem 6.7

(Phelps [40]) Let X be a Hausdorff space, $C(X)$ the bounded continuous complex functions on X. Then the (uniformly) closed unit ball B of $C(X)$ is the closed convex hull of the set U of functions in B of constant unit modulus.

Remarks

There is a far-reaching non-commutative version of Theorem 6.7, due to Russo and Dye [44], which is central to the modern theory of Banach *-algebras. One avatar of it is that in any norm closed, adjoint closed algebra of (bounded) linear operators on a complex Hilbert space which contains the identity operator, the closed unit ball is the closed convex hull of the unitary operators in the algebra. The proof by Russo and Dye is neither elementary nor transparent. But recently an exceedingly simple and natural proof was found by L. Harris [26], [27]. As is to be expected, his idea simplifies still further in a commutative setting and there emerges the following proof of Theorem 6.7.

Harris' Proof of Theorem 6.7: It suffices to consider $f \in C(X)$ with $\|f\|_\infty < 1$. For any $z, u \in \mathbb{C}$ with $|z| < 1 = |u|$ the number $\dfrac{z-u}{1-\bar{z}u} = \dfrac{-1}{u} \cdot \dfrac{z-u}{\bar{z}-\bar{u}}$ has modulus 1. Therefore the function

$$f_u = \frac{f-u}{1-u\bar{f}}$$

belongs to $C(X)$ and has constant unit modulus, that is, $f_u \in U$. Since $|\bar{f}| < 1$, we can expand f_u thus:

$$f_u = (f-u) \sum_{n=0}^{\infty} u^n \bar{f}^n = \sum_{n=0}^{\infty} u^n \bar{f}^n f - \sum_{n=1}^{\infty} u^n \bar{f}^{n-1}$$

$$= f + (|f|^2 - 1) \sum_{n=1}^{\infty} u^n \bar{f}^{n-1}.$$

Take for u various roots of unity and average. If we write f_r for $f_{e^{2\pi i r}}$ (r positive rational), this gives

$$(*) \quad \frac{1}{m} \sum_{k=1}^{m} f_{k/m} - f = (|f|^2 - 1) \sum_{n=1}^{\infty} [\frac{1}{m} \sum_{k=1}^{m} (e^{\frac{2\pi ik}{m}})^n] \bar{f}^{n-1}.$$

Now the average $\frac{1}{m} \sum_{k=1}^{m} (e^{\frac{2\pi in}{m}})^k$ is 1 or 0 according as $e^{\frac{2\pi in}{m}}$ is 1 or not, that is, according as m divides n or not. Thus in particular the first $m-1$ terms on the right of (*) are 0 and we get the following crude but quite adequate estimate

$$|\frac{1}{m} \sum_{k=1}^{m} f_{k/m} - f| \le ||f|^2-1| \sum_{n=m}^{\infty} |\bar{f}^{n-1}| = (1-|f|^2)|f|^{m-1} \sum_{n=1}^{\infty} |f|^n$$

$$= \frac{(1-|f|^2)|f|^{m-1}}{1 - |f|} = (1+|f|)|f|^{m-1} \le 2\|f\|_{\infty}^{m-1}.$$

As $\|f\|_{\infty} < 1$, it follows that f is the uniform limit of the convex sums $\frac{1}{m} \sum_{k=1}^{m} f_{k/m}$ and the proof is complete.

Remarks

Because $\|f\|_{\infty} < 1$, a simple calculation shows that $f_u(X) \subset \mathbb{C}\backslash\{ru: r \ge 0\}$. Now there exists an analytic logarithm

L_u in this latter region and so $f_u = e^{ig_u}$ where $g_u = -iL_u \circ f_u \in C(X)$ and in fact $e^{-\text{Im } g_u} = e^{\text{Re}(ig_u)} = |e^{ig_u}| = |f_u| = 1$ so $\text{Im } g_u = 0$ and $g_u \in C_{\mathbb{R}}(X)$. Therefore we have proved more than was claimed, namely that the unit ball of $C(X)$ is the closed convex hull of the functions e^{ig} with $g \in C_{\mathbb{R}}(X)$. In fact Phelps' original proof also establishes this. Badé & Curtis [3] give a proof of this improved form of Theorem 6.7 too. For yet another proof, see Sine [48].

Recalling that a point in a convex set is extreme if it is not the midpoint of two distinct points of the set, and checking the elementary fact that the set of extreme points of B is just U, we see the geometric significance of $B = \overline{\text{co}}\ U$. (Compare also the Krein-Milman and Choquet Theorems.) It is interesting that the corresponding result for $C_{\mathbb{R}}(X)$ does not hold for all compact Hausdorff X. Here again a function is extreme in the unit ball if and only if it has constant unit modulus, that is, takes only the values 1 and -1. The existence of such a function means a disconnection of X and such functions will be abundant enough to recover the whole unit ball by (limits of) their convex combinations only if X has lots of disconnection. The theorem, due to Badé (see [2]), is that the closed unit ball of $C_{\mathbb{R}}(X)$ (X compact Hausdorff) is the closed convex hull of its extreme points if and only if X is totally disconnected (that is, the only connected subsets of X are single points). See Goodner [23]. However Cantwell [13] has shown that for every $n \geq 2$ the unit ball in the space of bounded continuous $\mathbb{R}^n$-valued functions on X is the closed convex hull of its extreme points. We present the case $n = 2$ of his clever proof below. In this same vein Fisher [18] (cf. also Rudin [42]) shows that the closed unit ball in the disk algebra of continuous functions on the unit circle admitting analytic extensions into the unit disk is the closed convex hull of its unimodular functions; the latter, it is not hard to show, are the finite Blaschke products and each is an extreme point in the unit ball.

Let $D = \{z \in \mathbb{C}: |z| \le 1\}$, $B(x,\epsilon) = \{z \in \mathbb{C}: |z - x| < \epsilon\}$ for each $x \in \mathbb{C}$, $\epsilon \in \mathbb{R}$.

<u>Lemma 6.8</u>

Let X be a Hausdorff space and f a continuous function on X such that $f(X) \subset D\backslash\{y\}$ for some $y \in \mathbb{C}$ with $|y| < 1$. Then there exist continuous functions g,h on X with $g(X) \cup h(X) \subset D\backslash B(y,1-|y|)$ and $f = \frac{1}{2}(g + h)$.

<u>Proof</u>: Define a function λ on X by

$$\lambda(x) = \begin{cases} \sqrt{(1-|y|)^2 - |f(x)-y|^2} & \text{if } x \in f^{-1}(B(y,1-|y|)) \\ 0 & \text{if } x \in X\backslash f^{-1}(B(y,1-|y|)). \end{cases}$$

Evidently λ is continuous. So also then are the functions g and h defined on X by

$$g(x) = f(x) + i\lambda(x)\frac{f(x)-y}{|f(x)-y|}, \quad h(x) = f(x) - i\lambda(x)\frac{f(x)-y}{|f(x)-y|}.$$

Clearly $f = \frac{1}{2}(g + h)$. If $x \in X\backslash f^{-1}(B(y,1-|y|))$ then $g(x) = h(x) = f(x) \in f(X)\backslash B(y,1-|y|) \subset D\backslash B(y,1-|y|)$, while if $x \in f^{-1}(B(y,1-|y|))$ then

$$|g(x)-y| = |[f(x)-y] + \frac{i\sqrt{(1-|y|)^2 - |f(x)-y|^2}}{|f(x) - y|}[f(x)-y]|$$

$$= |f(x)-y|\ |1 + \frac{i\sqrt{(1-|y|)^2 - |f(x)-y|^2}}{|f(x) - y|}| = 1-|y|$$

and similarly for $h(x)$. Therefore g,h map all of X into $D\backslash B(y,1-|y|)$ [recall, this is an open ball].

Cantwell's Proof of Theorem 6.7 establishes a little more, namely that if $f \in C(X)$ and $\|f\|_\infty < 1$ then $f \in \text{co}\ U$. Given such an f let $\epsilon = 1 - \|f\|_\infty > 0$ so that

(1) $f(X) \subset \overline{B}(0,1-\epsilon) \subset D\backslash B(1,\epsilon)$.

Pick positive integer $m > 2/\epsilon$ and set

(2) $\begin{cases} y_1 = 1 \\ y_k = 1 - \frac{k}{m} \end{cases} \quad 2 \le k \le m.$

Then $y_2 \in B(y_1,\epsilon)$ so by (1) $y_2 \notin f(X)$ and by Lemma 6.8

(3.2) $\exists f_{2,j} \in C(X),\ f_{2,j}(X) \subset D\backslash B(y_2,1-|y_2|) \quad j = 1,2$

such that

(4.2) $f = \frac{1}{2}(f_{2,1} + f_{2,2})$.

Proceed inductively. Suppose for some $2 \le k < m$

(3.k) $\exists f_{k,j} \in C(X),\ f_{k,j}(X) \subset D\backslash B(y_k,1-|y_k|) \quad j = 1,2,\ldots,2^{k-1}$

such that

(4.k) $f = \frac{1}{2^{k-1}} \sum_{j=1}^{2^{k-1}} f_{k,j}$.

Then $|y_{k+1} - y_k| = \frac{1}{m} < \frac{k}{m} = 1 - |y_k|$ so $y_{k+1} \in B(y_k,1-|y_k|)$ and by (3.k) then $y_{k+1} \notin f_{k,j}(X)$. Therefore by 2^{k-1} applications of lemma 6.8 we see

(3.k+1) $\exists f_{k+1,j} \in C(X), f_{k+1,j}(X) \subset D\backslash B(y_{k+1}, 1-|y_{k+1}|)$ $j=1,2,\ldots,2^k$

such that

$$f_{k,j} = \tfrac{1}{2}(f_{k+1,2j-1} + f_{k+1,2j}) \qquad j = 1,2,\ldots,2^{k-1}$$

and consequently

(4.k+1) $$f = \frac{1}{2^{k-1}} \sum_{j=1}^{2^{k-1}} f_{k,j} = \frac{1}{2^k} \sum_{j=1}^{2^{k-1}} (f_{k+1,2j-1} + f_{k+1,2j})$$

Now $y_m = 0$ and so $B(y_m, 1-|y_m|) = B(0,1)$ and $D\backslash B(y_m, 1-|y_m|) = D\backslash B(0,1)$ is the unit circle. Therefore for the functions $f_{m,j}$ we have by (3.m) that $|f_{m,j}| \equiv 1$ so $f_{m,j} \in U$ and (4.m) gives $f = \frac{1}{2^{m-1}} \sum_{j=1}^{2^{m-1}} f_{m,j} \in \text{co } U$.

<u>Corollary 6.9</u>

If G is a locally compact abelian group and X a compact subset of $\hat{G}$, then $L^1(G)^\wedge|X = C(X)$ (that is, X is a Helson set) if there exist constants $k < 1$ and K such that for every $F \in C(X)$ with $|F| \equiv 1$ there is an $f \in L^1(G)$ with $\|f\|_1 \leq K$ and

$$\sup_{x \in X} |\hat{f}(x) - F(x)| \leq k.$$

<u>Proof</u>: Apply Theorem 6.5 to the map $T: L^1(G) \to C(X)$ given by $T(f) = \hat{f}|X$, taking E equal to the U of Theorem 6.7.

Next consider a locally compact Hausdorff space X, $C_0(X)$ the continuous complex functions vanishing at infinity on X, B the closed unit ball in $C_0(X)$ and B^+ the non-negative functions in B.

Lemma 6.10

If E is a bounded normal family in $C_0(X)$ then $\overline{co}\,E$ contains B^+.

Proof: It suffices to show $\overline{co}\,E$ contains every continuous $f:X \to [0,1)$ with compact support. Let such an f be given. Then $\|f\|_\infty < 1$. Let U be an open set with compact closure such that the closed support of f lies in U:

(1) $\overline{f^{-1}(0,1)} \subset U \text{ open} \subset \overline{U} \text{ compact}$

and consider any $n > [1 - \|f\|_\infty]^{-1}$, so that

(2) $$f(X) \subset [0,\tfrac{n-1}{n}).$$

For each positive integer $k = 1,2,\ldots,n-1$ set

$$U_k = \{x \in X: f(x) \geq k/n\}$$
$$V_k = \{x \in X: f(x) \leq (k-1)/n\} \cap \overline{U}.$$

Notice that

(3) $U_k \subset U,\ V_k \subset \overline{U}$

(4) $U_k \cap V_k = \emptyset$

for $k = 1,2,\ldots,n-1$.

Thus E, being normal, contains an f_1 such that

(5) $$f_1(U_1) = 1,\ f_1(V_1) = 0.$$

Set

(6) $W_1 = \{x \in X: |f_1(x)| \geq 1/n^2\} \cap (X\backslash U)$,

a compact set since $f_1 \in C_0(X)$. It is disjoint from U_2 by (3). Therefore $U_2, V_2 \cup W_1$ are disjoint compacta and there exists $f_2 \in E$ such that

$$f_2(U_2) = 1,\ f_2(V_2 \cup W_1) = 0.$$

Let

$$W_2 = \{x \in X: |f_2(x)| \geq 1/n^2\} \cap (X\setminus U),$$

etc. Continuing we finally obtain $f_1, f_2, \ldots, f_{n-1} \in E$ and associated sets $W_1, W_2, \ldots, W_{n-1}$ such that

$$(7) \qquad f_k(x) = \begin{cases} 1 & x \in U_k \\ 0 & x \in V_k \cup W_1 \cup \ldots \cup W_{k-1} \end{cases} \quad k = 2, \ldots, n-1.$$

$$(8) \qquad W_k = \{x \in X: |f_k(x)| \geq 1/n^2\} \cap (X\setminus U).$$

Notice that f_k is 0 in W_j for $j < k$ while $|f_k| \geq 1/n^2$ in W_k so

$$(9) \qquad W_j \cap W_k = \emptyset \quad \text{if} \quad j \neq k.$$

Define

$$(10) \qquad g = \frac{1}{n}\sum_{j=1}^{n-1} f_j \in \frac{n-1}{n} \text{ co } E.$$

We estimate $\|f - g\|_\infty$:

(I) If $x \in U$.

Then (recalling (2)) pick $k \in \{1,2,\ldots,n-1\}$ such that $\frac{k-1}{n} \leq f(x) \leq \frac{k}{n}$. It follows that $x \in U_1 \cap U_2 \cap \ldots \cap U_{k-1} \cap V_{k+1} \cap V_{k+2} \cap \ldots \cap V_{n-1}$

so from (7)

$$f_1(x) = f_2(x) = \ldots = f_{k-1}(x) = 1 \quad \text{and}$$

$$f_{k+1}(x) = f_{k+2}(x) = \ldots = f_{n-1}(x) = 0.$$

We therefore have

$$|f(x)-g(x)| \leq |f(x) - \frac{1}{n}\sum_{j=1}^{k-1} f_j(x)| + \frac{1}{n}|f_k(x)| + |\frac{1}{n}\sum_{j=k+1}^{n-1} f_j(x)|$$

$$= |f(x) - \frac{k-1}{n}| + \frac{1}{n}|f_k(x)|$$

$$\leq \frac{1}{n} + \frac{1}{n}\|f_k\|_\infty$$

(11) $\leq \frac{1}{n}(1 + M)$

where M is some bound on E.

(II) If $x \in (X\backslash U)\backslash \bigcup_{k=1}^{n-1} W_k$.

Then by (8) we have $|f_k(x)| < 1/n^2$ for $k = 1,2,\ldots,n-1$ and so $|g(x)| < \frac{1}{n} \sum_{k=1}^{n-1} \frac{1}{n^2} < \frac{1}{n^2}$ while $f(x) = 0$, since f is supported in U. Thus

(12) $|f(x) - g(x)| < \frac{1}{n^2}$.

(III) If $x \in (X\backslash U) \cap \sum_{k=1}^{n-1} W_k$.

Say $x \in W_k$. Then $x \in X\backslash U\backslash W_j$ for $j \neq k$ whence by (8) $|f_j(x)| < 1/n^2$. Again $f(x) = 0$ so

$$|f(x)-g(x)| = |g(x)| \leq \frac{1}{n}|f_k(x)| + \frac{1}{n} \sum_{j\neq k}^{n-1} |f_j(x)|$$

$$< \frac{1}{n}\|f_k\|_\infty + \frac{1}{n} \sum_{j\neq k}^{n-1} \frac{1}{n^2}$$

(13) $$< \frac{1}{n}M + \frac{1}{n^2}.$$

Combining (11), (12), (13), we get

$$|f(x) - g(x)| \leq \frac{1}{n}(M + 1) \qquad \forall x \in X.$$

(14) $\|f - g\|_\infty \leq \frac{1}{n}(M + 1)$.

Now

$$\|f - \tfrac{n}{n-1}g\|_\infty \leq \|f - g\|_\infty + \frac{1}{n-1}\|g\|_\infty$$

$$\leq \frac{1}{n}(M + 1) + \frac{1}{n-1}\,\frac{1}{n} \sum_{j=1}^{n-1} \|f_j\|_\infty \quad \text{by (10) \& (14)}$$

$$\leq \frac{1}{n}(M + 1) + \frac{1}{n-1}\,\frac{1}{n} \sum_{j=1}^{n-1} M$$

$$= \frac{1}{n}(2M + 1).$$

As $\frac{n}{n-1}g \in \text{co } E$ (10) and n may be chosen arbitrarily large, we see that $f \in \overline{\text{co}}\ E$.

Theorem 6.11

(Badé & Curtis [3]) Let X be a locally compact Hausdorff space, Y a Banach space continuously embedded in $C_0(X)$. If Y is boundedly ϵ-normal for some $\epsilon < \frac{1}{4}$, then $Y = C_0(X)$. If moreover K is a bound as in part (iv) of Definition 6.1, then $\| \ \|_Y \le 4K(1 - 4\epsilon)^{-1} \| \ \|_\infty$.

Proof: Let B denote the closed unit ball in $C_0(X)$. If F_1 and F_2 are disjoint compact subsets of X pick $h = h_{F_1,F_2} \in Y$ with

(1) $\|h\|_Y \le K$, $|h(F_1)-1| \le \epsilon$, $|h(F_2)| \le \epsilon$

and an $f = f_{F_1,F_2} \in B$ with

(2) $f(F_1) = 1$, $f(F_2) = 0$.

Then

$$\sup_{x \in F_1 \cup F_2} |f(x) - h(x)| \le \epsilon.$$

Let (Tietze) $g = g_{F_1,F_2} \in C_0(X)$ be an extension to X of $(f - h)|F_1 \cup F_2$ with

(3) $\|g\|_\infty \le \epsilon$.

We have $g + h = f$ on $F_1 \cup F_2$ so by (2)

(4) $(g + h)(F_1) = 1$, $(g + h)(F_2) = 0$.

Let A be the set consisting of all such functions $g_{F_1,F_2} + h_{F_1,F_2}$. Then A is uniformly bounded:

$$\|g + h\|_\infty \le \|g\|_\infty + \|h\|_\infty \le \|g\|_\infty + M\|h\|_Y \le \epsilon + MK < \tfrac{1}{4} + MK,$$

where we have taken for M the norm of the injection of Y

into $C_0(X)$. Moreover from (4) we see that A is a normal family. From Lemma 6.10 then $\overline{\text{co}}\ A \supset B^+$ and so obviously

(5) $\overline{\text{coe}}\ 4A \supset B.$

Since by (3)

$$\|4(g + h) - 4h\|_\infty = 4\|g\|_\infty$$

(6) $$\leq 4\epsilon,$$

we are in a position to apply Theorem 6.5 with T the injection map of Y into $C_0(X)$, $E = 4A$, $k = 4\epsilon$ and $c = 4K$. Because of (6), (1) and (5) the hypotheses of Theorem 6.5 are met and we conclude that the range Y of T is all of $C_0(X)$ and the norm inequality holds.

Note

If Y is a Banach algebra, lemma 6.2 allows the hypothesis in the last theorem and in Corollary 6.13 below to be relaxed to $\epsilon < \frac{1}{2}$. Moreover the continuity of the embedding need not be hypothesized because $\| \ \|_\infty \leq \| \ \|_Y$ is automatic (III, p. 1).

Corollary 6.12

If X is a compact Hausdorff space and A a Banach algebra lying in $C(X)$ which is normal and for which there exists a constant M such that the idempotents in each quotient algebra A/kF (F closed $\subset X$) are bounded by M, then $A = C(X)$ and $\| \ \|_A \leq 4M\| \ \|_\infty$.

Proof: If E_1, E_2 are disjoint closed subsets of X, then normality of A provides a function $f \in A$ such that

(1) $f(E_1) = 1 - f(E_2) = 0.$

So $f + k(E_1 \cup E_2)$ is an idempotent in $A/k(E_1 \cup E_2)$ and hence has (quotient) norm no greater than M. By definition of the quotient norm this means that, given $\delta > 0$, the $f \in A$ can be chosen to satisfy in addition to (1)

(2) $$\|f\|_A < M + \delta.$$

We have shown that for each $\delta > 0$, A is boundedly ϵ-normal with bound $M + \delta$ for $\epsilon = 0$. All conclusions therefore follow from the last theorem.

Corollary 6.13

Let X be a compact Hausdorff space, Y a Banach space continuously embedded in C(X). If for some $\epsilon < \frac{1}{4}$, Y is locally boundedly ϵ-normal, then X can be covered by finitely many open sets U_j such that $Y|\overline{U}_j = C(\overline{U}_j)$, $j = 1,2,\ldots,n$.

Proof: From the open sets U provided by the definition 6.1(v) of locally boundedly ϵ-normal, extract a finite cover $U_1,\ldots,U_n$ of X. Consider $Y_j = Y|\overline{U}_j$ in the quotient (pseudo-) norm of $Y/k\overline{U}_j$, $k\overline{U}_j$ the vector subspace of functions in Y zero on $\overline{U}_j$. By hypothesis there is a constant M such that $\| \;\|_\infty \le M\| \;\|_Y$. This implies that each $k\overline{U}_j$ is closed in Y so Y_j is a Banach space. This inequality also implies that $\| \;\|_\infty \le M\| \;\|_{Y_j}$ in Y_j. See (IV), p. 2. It is also clear upon a little reflection that Y_j is a boundedly ϵ-normal family in $C(\overline{U}_j)$. It follows from Theorem 6.11 then that $Y_j = C(\overline{U}_j)$.

When Y is an algebra we can make a (not entirely routine)

partition of unity argument and conclude $Y = C(X)$ (Corollary 6.15 below).

Lemma 6.14

Let X be a compact Hausdorff space, A a Banach algebra lying in $C(X)$ which is ϵ_0-normal for some $\epsilon_0 < \frac{1}{2}$. Suppose X is covered by open sets $U_1,\ldots,U_n$ such that $A|\overline{U}_k = C(\overline{U}_k)$. Then for each $\epsilon > 0$ there exists $f_k \in A$ such that

(i) $\quad |f_k(X\backslash U_k)| \le \epsilon \qquad k = 1,2,\ldots,n$

(ii) $\quad f_1 + \ldots + f_n = 1.$

Proof: Let us call such a set of functions $\{f_1,\ldots,f_n\}$ an ϵ-partition of unity subordinate to the cover $\{U_1,\ldots,U_n\}$. Since $A|\overline{U}_k = C(\overline{U}_k)$ and $\|\ \|_\infty \le \|\ \|_A$, whence, as noted before the quotient norm in $A|\overline{U}_k$ dominates the uniform norm in $C(\overline{U}_k)$, the Open Map Theorem provides a constant C_k such that

(1) $\quad f \in C(\overline{U}_k) \Rightarrow f = \tilde{f}|\overline{U}_k$ for some $\tilde{f} \in A$ with $\|\tilde{f}\|_A \le C_k\|f\|_\infty$.

We set

(2) $$C = \max_{1\le k\le n} C_k$$

and let U be some U_k. Let $0 < \delta < \frac{1}{2}$ be given and any closed $S \subset U$. According to Remark 6.3 and our hypothesis, A is $\delta/2C+1$ normal, so there exists $h \in A$ with

(3) $$\begin{cases} |1 - h(S)| \le \delta/2C+1 \\ |h(X\backslash U)| \le \delta/2C+1. \end{cases}$$

Let $T = \{x \in X: |h(x)| \ge \frac{1}{2}\}$. Because of (3) and $\delta < \frac{1}{2}$ we have

(4) $\qquad S \subset T \subset U.$

Tietze insures that the function $\frac{1}{h}|T \in C(T)$ has an extension to $g \in C(\overline{U})$ which satisfies

$$\|g\|_\infty = \sup_T |\tfrac{1}{h}| \le 2.$$

According to (1) (and (2)) therefore there is a $\tilde{g} \in A$ with

(5) $\qquad \|\tilde{g}\|_A \le C\|g\|_\infty \le 2C$

$\qquad \tilde{g}|\overline{U} = g.$

We set $\varphi = \tilde{g}h \in A$ and have

(6) $\qquad \varphi(S) \subset \varphi(T) = 1 \quad$ (since $\tilde{g}$ extends $\frac{1}{h}|T$)

while

(7) $\qquad |\varphi(X\backslash U)| \le \|\tilde{g}\|_\infty \, |h(X\backslash U)| \overset{(3)}{\le} \|\tilde{g}\|_A \frac{\delta}{2C+1} \overset{(5)}{\le} 2C\cdot \frac{\delta}{2C+1} < \delta$

and by definition of T

(8) $\qquad |\varphi(U\backslash T)| \le \|\tilde{g}\|_\infty \, |h(U\backslash T)| \le \|\tilde{g}\|_A \cdot \tfrac{1}{2} \overset{(5)}{\le} C.$

In particular from (6), (7), (8) [upon recalling that $C \ge 1$] we have

(9) $\qquad \|\varphi\|_\infty \le C.$

Summarily

(10) $\quad S$ closed $\subset U_k, 0<\delta<\tfrac{1}{2} \Rightarrow \exists\varphi \in A$ with $\varphi(S)=1, |\varphi(X\backslash U_k)|<\delta, \|\varphi\|_\infty \le C.$

Now choose V_k open $\subset \overline{V}_k \subset U_k$ so that $\{V_1,\ldots,V_n\}$ covers X. This can be done successively thus: $X\backslash U_2 \cup \ldots \cup U_n$ is a closed subset of U_1. So there exists an open set V_1 with $X\backslash U_2 \cup \ldots \cup U_n \subset V_1 \subset \overline{V}_1 \subset U_1$. Then $V_1, U_2, \ldots, U_n$ cover X. With open sets $V_1,\ldots,V_k$ chosen $(1 \le k < n)$ so that $V_1,\ldots,V_k,U_{k+1},\ldots,U_n$ cover X and $\overline{V}_j \subset U_j$ $(1 \le j \le k)$, the set $X\backslash V_1 \cup \ldots \cup V_k \cup U_{k+2} \cup \ldots \cup U_n$ is closed and lies in U_{k+1}. The choice of V_{k+1} is now clear. Use (10) to choose successively $\varphi_n, \ldots, \varphi_1 \in A$ such that

(11) $$|\varphi_k(X\backslash U_k)| \le \epsilon[\|\prod_{j=k+1}^{n}(1 - \varphi_j)\|_\infty + 1]^{-1}$$

(12) $$\varphi_k(\overline{V}_k) = 1,$$

where a void product means 1. Set

(13) $$f_k = \varphi_k \prod_{j=k+1}^{n}(1 - \varphi_j) \in A.$$

We have from (12) and the fact $X = V_1 \cup \ldots \cup V_n$ that

(14) $$\prod_{k=1}^{n}(1 - \varphi_k) = 0.$$

But for any complex numbers $\lambda_1, \ldots, \lambda_n$ a trivial induction establishes that

(15) $$\prod_{k=1}^{n}(1 - \lambda_k) = 1 - \lambda_n - \sum_{k=1}^{n-1}\lambda_k \prod_{j=k+1}^{n}(1 - \lambda_j).$$

Therefore from (14)

(16) $$1 = \varphi_n + \sum_{k=1}^{n-1}\varphi_k \prod_{j=k+1}^{n}(1 - \varphi_j) = f_n + \sum_{k=1}^{n-1} f_k.$$

This is desideratum (ii). Finally from (11) and (13) follow (i).

Corollary 6.15

Let X be a compact Hausdorff space, A a Banach algebra lying in $C(X)$ which is ϵ_0-normal for some $\epsilon_0 < \frac{1}{2}$. Suppose X is covered by open sets $U_1, \ldots, U_n$ such that $A|\overline{U}_k = C(\overline{U}_k)$. Then $A = C(X)$.

Proof: Let C be the constant (2) defined in the last proof and set $\epsilon = (2n(C+1))^{-1}$. For this ϵ let $f_1,\ldots,f_n$ be as provided by the last lemma. We have in mind applying Theorem 6.5 with E the unit ball of $C(X)$, $k = \frac{1}{2}$, $c = C\sum_{j=1}^{n}\|f_j\|_A$, and T injection. For if $g \in E$ is given we may, because of how C is defined, find $g_1,\ldots,g_n \in A$ satisfying

$$g_j|\overline{U}_j = g|\overline{U}_j$$

and

$$(1) \qquad \|g_j\|_A \le C\|g|\overline{U}_j\|_\infty \le C\|g\|_\infty \le C.$$

Set

$$(2) \qquad f = \sum_{j=1}^{n} g_j f_j \in A.$$

Then

$$(3) \qquad \|f\|_A \le \sum_{j=1}^{n}\|g_j\|_A\|f_j\|_A \overset{(1)}{\le} C\sum_{j=1}^{n}\|f_j\|_A = c.$$

On the other hand

$$gf_j = g_jf_j \quad \text{in } U_j \quad \text{and}$$

$$|gf_j - g_jf_j| \le (\|g\|_\infty + \|g_j\|_\infty)|f_j| \le (1+C)|f_j| \le (1+C)\epsilon \quad \text{in } X\backslash U_j$$

by property (i) of the f_j. Thus

$$(4) \qquad \|gf_j - g_jf_j\|_\infty \le (1+C)\epsilon.$$

It follows (recalling (ii) and (2)) that

$$(5) \qquad \|g-f\|_\infty = \|\sum_{j=1}^{n} gf_j - \sum_{j=1}^{n} g_jf_j\|_\infty$$

$$\le \sum_{j=1}^{n}\|gf_j - g_jf_j\|_\infty \le n(1+C)\epsilon$$

$$= \tfrac{1}{2} \text{ by definition of } \epsilon.$$

Since (3) and (5) hold for each $g \in E$, the unit ball of $C(X)$, Theorem 6.5 applies and we conclude $A = C(X)$.

Corollary 6.16

(Badé & Curtis [3]) Let X be a compact Hausdorff space, A a Banach algebra lying in $C(X)$ which is locally boundedly ϵ-normal for some $\epsilon < \frac{1}{2}$. Then $A = C(X)$.

Proof: Quote Corollary 6.15 and Corollary 6.13 (plus the note to Theorem 6.11).

Remarks

In connection with Corollary 6.16, it should be noted that, even for uniformly closed subalgebras A, normality without some boundedness hypothesis is not enough to conclude $A = C(X)$. McKissick [36] and [37] has constructed a compact subset X of $\mathbb{C}$ and a uniformly closed proper subalgebra A of $C(X)$ which is normal. In fact, in his example A is the uniform closure in $C(X)$ of the rational functions with poles outside X. There is a detailed exposition of McKissick's example in [49], pp. 346-355.

Chapter VII

KATZNELSON'S BOUNDED IDEMPOTENT THEOREM

Our goal here is to relax the requirement in Corollary 6.12 that the bound on the idempotents in the quotient algebras be uniform. We need some definitions and lemmas. The setting for the next three lemmas (from Katznelson [32]) is a compact Hausdorff space X and a Banach algebra A lying in $C(X)$ which is normal and for which the idempotents in each quotient algebra A/kF are bounded. We will eventually prove (Theorem 7.8 below) that A must be all of $C(X)$.

Definition 7.1

Let us say $\underline{A}$ is bounded on a subset $\underline{V}$ of $\underline{X}$ if there exists a constant B_V such that the idempotents in A/kF are all bounded (quotient norm) by B_V whenever F is a closed subset of V. Say $\underline{A}$ is bounded at a point $\underline{x}$ of $\underline{X}$ if A is bounded on some neighborhood of x.

Lemma 7.2

If A is bounded on each of the open sets V_1, V_2 then A is bounded on every closed subset F of $V_1 \cup V_2$.

Proof: Since $F \backslash V_2$ is a closed subset of open V_1 there exists open W_1 such that

$$F \backslash V_2 \subset W_1 \subset \overline{W}_1 \subset V_1.$$

Then similarly there exists open W_2 such that

$$F\backslash W_1 \subset W_2 \subset \overline{W}_2 \subset V_2.$$

As $F\backslash W_1$, $F\backslash W_2$ are disjoint closed subsets of X, and A is normal, there is a $\varphi \in A$ with

(1) $$\varphi(F\backslash W_1) = 0, \ \varphi(F\backslash W_2) = 1.$$

Now let P be any closed subset of F and let $f + kP$ be any idempotent in A/kP. This means that f is 0 or 1 at every point of P and so f is 0 or 1 at every point of $P \cap \overline{W_i}$ $(i = 1,2)$, which means that $f + k(P \cap \overline{W_i})$ $(i = 1,2)$ is an idempotent. Therefore, as the V_i are sets on which A is bounded and $P \cap \overline{W_i}$ is a closed subset of V_i,

$$\|f + k(P \cap \overline{W_i})\| \leq B_{V_i}.$$

By definition of the quotient norm then there exist $f_i \in f+k(P\cap\overline{W}_i)$ such that

(2) $$\|f_i\|_A < B_{V_i} + 1.$$

Now $f_1 \in f + k(P \cap \overline{W}_1)$ means $f_1 = f$ on $P \cap \overline{W}_1$. Therefore

(3) $$\varphi f = \varphi f_1 \quad \text{on} \quad P \cap \overline{W}_1.$$

Of course since $P \subset F$ we have

$$P\backslash(P \cap \overline{W}_1) \subset F\backslash\overline{W}_1$$

and so by (1) we have that φ is 0 on $P\backslash(P \cap \overline{W}_1)$ so

(4) $$\varphi f = \varphi f_1 \quad \text{on} \quad P\backslash(P \cap \overline{W}_1).$$

From (3) and (4)

$$\varphi f = \varphi f_1 \quad \text{on} \quad P.$$

Similarly $(1 - \varphi)f = (1 - \varphi)f_2$ on P and so

$$f = \varphi f + (1-\varphi)f = \varphi f_1 + (1-\varphi)f_2 \quad \text{on} \quad P,$$

that is, $\varphi f_1 + (1-\varphi)f_2 \in f + kP$. It follows that

$$\|f+kP\| \le \|\varphi f_1 + (1-\varphi)f_2\|_A \le \|\varphi\|_A\|f_1\|_A + \|1-\varphi\|_A\|f_2\|_A$$

$$(5) \qquad \le \|\varphi\|_A(B_{V_1} + 1) + \|1-\varphi\|_A(B_{V_2} + 1) \quad \text{by (2).}$$

Therefore the right side of (5) may be taken as a constant B_F which bounds all idempotents in A/kP for every closed $P \subset F$.

Lemma 7.3

If A is bounded on each of the open sets $V_1,\ldots,V_n$ then A is bounded on every closed subset of $V_1 \cup \ldots \cup V_n$.

Proof: By induction on n. Trivial for $n = 1$. If true for $n = k$ and A is bounded on each of the open sets $V_1,\ldots,V_k,V_{k+1}$ and F is a closed subset of $V_1 \cup \ldots \cup V_k \cup V_{k+1}$, then $F\backslash V_{k+1}$ is a closed subset of $V_1 \cup \ldots \cup V_k$. Pick open U such that

$$F\backslash V_{k+1} \subset U \subset \overline{U} \subset V_1 \cup \ldots \cup V_k.$$

By the induction hypothesis, A is bounded on $\overline{U}$, consequently on U also, and therefore by lemma 7.2 on the closed subset F of $U \cup V_{k+1}$.

Corollary 7.4

If F is closed and A is bounded at every point of F, then A is bounded on some open neighborhood V of F.

Proof: Immediate from the definition of A bounded at a point, the last lemma, and the compactness of $F: F \subset V_1 \cup\ldots\cup V_n$ where each V_j is an open set in which A is bounded. Pick open V so that $F \subset V \subset \overline{V} \subset V_1 \cup \ldots \cup V_n$. By lemma 7.3,

A is bounded on $\overline{V}$ and so on V.

Lemma 7.5

There are at most finitely many $x \in X$ at which A is not bounded.

Proof: As in the proof of lemma 4.22, if the conclusion fails there exist mutually disjoint open U_n on which A is not bounded. A not bounded on U_n means there exist closed $F_n \subset U_n$ and idempotent $f_n + kF_n$ in A/kF_n with

(1) $$\|f_n + kF_n\| \geq n.$$

Now let $F = (\bigcup_{n=1}^{\infty} F_n)^-$. Since $U_n \cap F_m = \emptyset$ if $n \neq m$, $F = F_n \cup (\bigcup_{m\neq n} F_m)^-$ and $F_n \cap (\bigcup_{m\neq n} F_m)^- = \emptyset$. Thus by normality there exist $\varphi_n \in A$ with

(2) $$\varphi_n(F_n) = 1, \quad \varphi_n(F\backslash F_n) = 0.$$

As $f_n + kF_n$ is idempotent we know f_n is 0 or 1 on F_n, so from (2) we get

(3) $\varphi_n f_n$ is 0 or 1 in F.

Yet $F \supset F_n$ clearly implies $kF \subset kF_n$ so that

(4) $$\|\varphi + kF_n\| \leq \|\varphi + kF\| \qquad \forall \varphi \in A$$

and we see

$$n \overset{(1)}{\leq} \|f_n + kF_n\| \overset{(2)}{=} \|\varphi_n f_n + kF_n\| \overset{(4)}{\leq} \|\varphi_n f_n + kF\|,$$

which says (recalling (3)) that the idempotents in A/kF are not bounded, contrary to the basic hypothesis about A.

Lemma 7.6

(Katznelson [33]) Let X be a compact Hausdorff space and B a Banach algebra lying in $C(X)$ with $1 \in B$. Suppose for some $x_0 \in X$ and some base $\mathcal{V}$ of open neighborhoods of x_0 there is a constant K such that for every $V \in \mathcal{V}$

(i) $$B_0|X\backslash V = C(X\backslash V)$$

(ii) $$\|f + k(X\backslash V)\|_{B/k(X\backslash V)} \le K\|f|X\backslash V\|_\infty \qquad \forall f \in B_0$$

where $B_0 = \{f \in B : f(x_0) = 0\}$. Then $B = C(X)$.

Proof: Since $1 \in B$ it suffices to show that B contains each $g \in C(X)$ with $g(x_0) = 0$. We may also assume $g \neq 0$. Choose $V_1 \in \mathcal{V}$ such that $g|X\backslash V_1 \neq 0$ and

(1) $$|g(x)| < \frac{1}{8(K+3)}\|g\|_\infty \qquad \forall x \in V_1.$$

By (i) there exists $h_1 \in C(X)$ such that $g + h_1 \in B_0$ and $h_1(X\backslash V_1) = 0$. By (ii) we can even choose h_1 so that

(2) $$\|g+h_1\|_B < (K+1)\|g|X\backslash V_1\|_\infty \le (K+1)\|g\|_\infty, \quad h_1(X\backslash V_1) = 0.$$

Choose W_1 such that

(3) $$\begin{cases} x_0 \in W_1 \text{ open} \subset \overline{W_1} \subset V_1 & \text{and} \\ |h_1(X\backslash W_1)| < \dfrac{\|g\|_\infty}{4}. \end{cases}$$

This is possible since $h_1(X\backslash V_1) = 0$ implies h_1 is small on some neighborhood of the compact set $X\backslash V_1$. Now g vanishes at x_0 and $g + h_1 \in B_0$ so $g + h_1$ vanishes at x_0. Therefore $h_1(x_0) = 0 = g(x_0)$ and we may choose $U_1 \in \mathcal{V}$ such that

(4) $$\begin{cases} x_0 \in U_1 \text{ open} \subset W_1 & \text{and} \\ |h_1(x)| + |g(x)| < \dfrac{1}{4(K+3)}\|g\|_\infty & \forall x \in U_1. \end{cases}$$

Take $\varphi_1 \in C(X)$ with

(5) $$\varphi_1(X) \subset [0,1],\ \varphi_1(X\backslash V_1) = 1,\ \varphi_1(\overline{W}_1) = 0.$$

Then use (i) and (ii) to pick $h_1' \in C(X)$ such that $\varphi_1 + h_1' \in B_0$ and

(6) $$\|\varphi_1+h_1'\|_B < (K+1)\|\varphi_1|X\backslash U_1\|_\infty \le K+1 \quad \text{and} \quad h_1'(X\backslash U_1) = 0.$$

Then

(7) $$\|h_1'\|_\infty \le \|\varphi_1\|_\infty + \|\varphi_1+h_1'\|_\infty \le 1 + \|\varphi_1+h_1'\|_B \le 2+K.$$

Set

$$g_1 = (\varphi_1 + h_1')(g + h_1) \in B_0.$$

Claim:

(8) $$\|g - g_1\|_\infty \le \tfrac{1}{2}\|g\|_\infty.$$

If $x \in X\backslash V_1$ then $h_1(x) = 0$ (by (2)), $h_1'(x) = 0$ (by (6)) and $\varphi_1(x) = 1$ (by (5)). Therefore

$$g_1(x) - g(x) = 0.$$

If $x \in V_1\backslash W_1$ then

$$\begin{aligned}
|g_1(x)-g(x)| &\le |g(x)| + [|\varphi_1(x)|+|h_1'(x)|][|g(x)|+|h_1(x)|] \\
&< \tfrac{1}{8}\|g\|_\infty + [|\varphi_1(x)|+|h_1'(x)|][\tfrac{1}{8}\|g\|_\infty+|h_1(x)|] \quad \text{by (1)} \\
&\le \tfrac{1}{8}\|g\|_\infty + [1+0][\tfrac{1}{8}\|g\|_\infty+|h_1(x)|] \quad \text{by (6) and (5)} \\
&\le \tfrac{1}{8}\|g\|_\infty + [\tfrac{1}{8}\|g\|_\infty + \tfrac{1}{4}\|g\|_\infty] \quad \text{by (3)} \\
&= \tfrac{1}{2}\|g\|_\infty.
\end{aligned}$$

If $x \in W_1\backslash U_1$ then $\varphi_1(x) = 0$ (by (5)) and $h_1'(x) = 0$ (by (6)). Therefore

$$|g_1(x) - g(x)| = |-g(x)| < \tfrac{1}{2}\|g\|_\infty \quad \text{by (1).}$$

If $x \in U_1$ then

$$|g_1(x)-g(x)| \le |g(x)|+|g_1(x)| \le \tfrac{1}{8}\|g\|_\infty + [|\varphi_1(x)+h_1'(x)|][|g(x)|+|h_1(x)|]$$

$$\le \tfrac{1}{8}\|g\|_\infty + [0+2+K][|g(x)|+|h_1(x)|] \quad \text{by (5) and (7)}$$

$$\le \tfrac{1}{8}\|g\|_\infty + [2+K]\frac{1}{4(2+K)}\|g\|_\infty \quad \text{by (4)}$$

$$< \tfrac{1}{2}\|g\|_\infty.$$

Thus (8) is established. Finally from (2) and (6) and the definition of g_1 we get

(9) $$\|g_1\|_B \le (K+1)^2\|g\|_\infty.$$

We are now in a position to quote Theorem 6.5. Alternatively we can iterate the construction with $g-g_1$ in the role of g etc. We get $g_1, g_2, g_3, \ldots \in B_0$ such that

(10) $$\|g - (g_1 + \ldots + g_n)\|_\infty \le \tfrac{1}{2}\|g - (g_1 + \ldots + g_{n-1})\|_\infty$$

(11) $$\|g_n\|_B \le (K+1)^2\|g - (g_1 + \ldots + g_{n-1})\|_\infty.$$

Induction on (10) gives

(12) $$\|g - (g_1 + \ldots + g_n)\|_\infty \le \frac{1}{2^n}\|g\|_\infty$$

and then (11) yields

(13) $$\|g_n\|_B \le \frac{(K+1)^2}{2^{n-1}}\|g\|_\infty.$$

The series $\sum_{n=1}^{\infty} g_n$ therefore converges in the norm of B to g.

Lemma 7.7

Let X be a compact Hausdorff space, A a Banach space lying in $C(X)$. Let V be a closed subset of X and B the space $A|V$ with the quotient norm of A/kV. Then for each closed $F \subset V$ the two quotient norms in $A|F$ are equal:

$$\|f|V + kF\|_{B/kF} = \|f + kF\|_{A/kF} \qquad \forall f \in A$$

where (suffering a little abuse of language) kF denotes the vector space of functions in A or B, as the case may be, which vanish on F.

Proof: Given $g \in A$ with $g(F) = 0$ we have

$$\|f|V + g|V\|_B = \|(f + g)|V\|_B \le \|f + g\|_A$$

as the norm in $B = A|V$ is the quotient norm. Therefore

$$\|f|V + kF\|_{B/kF} \le \|f|V + g|V\|_B \le \|f + g\|_A.$$

Taking the infimum over such g

(1) $\quad \|f|V + kF\|_{B/kF} \le \|f + kF\|_{A/kF}.$

Next given $\epsilon > 0$ and $g \in B$ with $g(F) = 0$ pick $h \in A$ such that $h|V = f|V + g$ and

$$\|h\|_A < \|f|V + g\|_B + \epsilon.$$

As $h|F = (f|V)|F + g|F = f|F$ we have $h \in f + kF$ so

$$\|f + kF\|_{A/kF} \le \|h\|_A < \|f|V + g\|_B + \epsilon.$$

Taking the infimum on such g in this last inequality gives

(2) $\quad \|f + kF\|_{A/kF} \le \|f|V + kF\|_{B/kF} + \epsilon$

and as $\epsilon > 0$ is arbitrary, (1) and (2) give the desired equality.

Theorem 7.8

(Katznelson [32] and [34]) Let X be a compact Hausdorff space and A a Banach algebra lying in $C(X)$ which is normal and for which the idempotents in each quotient algebra A/kF (F closed $\subset X$) are bounded. Then $A = C(X)$.

Proof: It suffices to show that A is bounded at each point of X. For then by Corollary 7.4 A is bounded on X and the desired conclusion is guaranteed by Corollary 6.12.

Toward this end we notice that as a consequence of lemma 7.5 each point x of X has a neighborhood V such that A is bounded at every point of $\overline{V}\backslash\{x\}$.

Next we notice that if E is a closed subset of X and $A|E$ is considered in the quotient norm of A/kE then $A|E$ satisfies the same hypotheses that A does: normality is trivial and for the boundedness of idempotents in quotients we just quote lemma 7.7.

Because of the last paragraph it suffices to deal in the second paragraph only with the case $\overline{V} = X$. Summarily then the theorem will be proved if, under the assumption that for some $x_0 \in X$, A is bounded at every point of $X\backslash\{x_0\}$, we can prove that A must also be bounded at x_0.

With a view to using lemma 7.6 we will next show that there is a closed neighborhood V of x_0 and a constant M with the property that the idempotents in $A/k(F \cup \{x_0\})$ are bounded by M for all compact $F \subset V\backslash\{x_0\}$. The argument for this is like that in the proof of lemma 7.5: we suppose it false and construct inductively closed neighborhoods V_n of x_0 and compact $F_n \subset V_n\backslash\{x_0\}$ such that $A/k(F_n \cup \{x_0\})$ contains an idempotent, say $g_n + k(F_n \cup \{x_0\})$ $(g_n \in A)$, of norm greater than $n+1$ and such that for $n > 1$, V_n is disjoint from $F_1, F_2, \ldots, F_{n-1}$. Now g_n takes only the values 0 and 1 on $F_n \cup \{x_0\}$ and so $1-g_n$ also has this property, that is, $1 - g_n + k(F_n \cup \{x_0\})$ is also an idempotent. Moreover its norm is greater than n:

$$\|1-g_n+k(F_n \cup \{x_0\})\| \geq \|g_n+k(F_n \cup \{x_0\})\| - \|1+k(F_n \cup \{x_0\})\|$$
$$\geq \|g_n + k(F_n \cup \{x_0\})\| - 1$$
$$> n+1 - 1 = n.$$

(Note that we have used the fact $1 \in \mathbf{A}$, an immediate consequence of normality.) Since one of g_n and $1-g_n$ must be 0 at x_0 we see that $A/k(F_n \cup \{x_0\})$ contains an idempotent $f_n + k(F_n \cup \{x_0\})$ ($f_n = g_n$ or $1-g_n$ as the case may be) with

(1) $$\|f_n + k(F_n \cup \{x_0\})\| > n$$

(2) $$f_n(x_0) = 0.$$

As in the proof of lemma 7.5 we set $F = \{x_0\} \cup (\bigcup_{n=1}^{\infty} F_n)^-$. We have that $F_m \subset V_{n+1}$ for $m \geq n+1$ so

$$(\bigcup_{m=n+1}^{\infty} F_m)^- \subset V_{n+1} \subset X \backslash F_1 \cup F_2 \cup \ldots \cup F_n$$

and as the F_j are all disjoint we see that

$$F_n \subset X \backslash [F_1 \cup F_2 \cup \ldots \cup F_{n-1} \cup (\bigcup_{m=n+1}^{\infty} F_m)^-] = X \backslash (\bigcup_{m \neq n}^{\infty} F_m)^-.$$

Therefore $F = F_n \cup (\bigcup_{m \neq n}^{\infty} F_m \cup \{x_0\})^-$ is a disjoint union and we can appeal to the normality of $\mathbf{A}$ for a function $\varphi_n \in \mathbf{A}$ with

(3) $$\varphi_n(F_n) = 1, \ \varphi_n(F \backslash F_n) = 0.$$

As f_n is 0 or 1 on F_n and 0 at x_0 we see from (3) that

(4) $$\varphi_n f_n + k(F_n \cup \{x_0\}) = f_n + k(F_n \cup \{x_0\})$$

and also from (3) that $\varphi_n f_n$ is 0 or 1 throughout F; so

(5) $\varphi_n f_n + kF$ is an idempotent.

Then, as before,

$$n \overset{(1)}{<} \|f_n + k(F_n \cup \{x_0\})\| \overset{(4)}{=} \|\varphi_n f_n + k(F_n \cup \{x_0\})\| \le \|\varphi_n f_n + kF\| ,$$

which says that the idempotents in A/kF are not bounded, contrary to the basic hypothesis on A.

With the neighborhood V and constant M in hand we assert that for every compact $F \subset V\backslash\{x_0\}$, $A|F \cup \{x_0\}$ (quotient norm) satisfies the hypotheses of Corollary 6.12. For if P closed $\subset F \cup \{x_0\}$ two cases must be looked at: If $x_0 \in P$ then $P = F_0 \cup \{x_0\}$ where $F_0 = P\backslash\{x_0\}$ is a compact subset of $V\backslash\{x_0\}$. Then the idempotents in $A/k(F_0 \cup \{x_0\}) = A/kP$ are bounded by M and so (lemma 7.7) those of the corresponding quotient of $A|F \cup \{x_0\}$ are also bounded by M. If $x_0 \notin P$ then P is a closed subset of $V\backslash\{x_0\}$ so the idempotents in $A/k(P \cup \{x_0\})$ are bounded by M. By normality A contains an h such that

$$h(P) = 1, \; h(x_0) = 0.$$

Then for any idempotent $f + kP$ in A/kP we shall have $hf + kP = f + kP$ and $hf + k(P \cup \{x_0\})$ is also an idempotent. It follows that

$$\|f + kP\| = \|hf + kP\| \le \|hf + k(P \cup \{x_0\})\| \le M.$$

That is, the idempotents in A/kP are bounded by M. By lemma 7.7 the idempotents in the corresponding quotient of $A|F \cup \{x_0\}$ are also bounded by M.

Because of the result of the last paragraph we can use Corollary 6.12 on $A|F \cup \{x_0\}$ for each closed $F \subset V\backslash\{x_0\}$ and get

(6) $\quad A|F \cup \{x_0\} = C(F \cup \{x_0\})$

(7) $\quad \|f + k(F \cup \{x_0\})\|_{A/k(F\cup\{x_0\})} \le 4M\|f|F \cup \{x_0\}\|_\infty \quad \forall f \in A.$

Now we have in mind applying lemma 7.6 to V in the role of X and the algebra $A|V$ (quotient norm) in the role of B. Setting $A_0 = \{f \in A: f(x_0) = 0\}$ and $B_0 = A_0|V$ it is obvious from (6) that

$$B_0|F = C(F) \quad \text{for all compact } F \subset V\backslash\{x_0\}$$

that is, if $\mathcal{V}$ is the set of all open neighborhoods of x_0 which lie in V,

(i) $\quad B_0|V\backslash U = C(V\backslash U) \quad$ for all $U \in \mathcal{V}$.

But from (7) we have for any closed $F \subset V\backslash\{x_0\}$

$$\begin{aligned} \|f + kF\|_{A/kF} &\le \|f + k(F \cup \{x_0\})\|_{A/k(F \cup \{x_0\})} \\ &\le 4M\|f|F \cup \{x_0\}\|_\infty \\ &= 4M\|f|F\|_\infty \qquad \forall f \in A_0. \end{aligned}$$

That is, for every $U \in \mathcal{V}$

(8) $\quad \|f + k(V\backslash U)\|_{A/k(V\backslash U)} \le 4M\|f|V\backslash U\|_\infty \qquad \forall f \in A_0.$

From this and lemma 7.7 it follows at once that

(ii) $\quad \|f + k(V\backslash U)\|_{B/k(V\backslash U)} \le 4M\|f|V\backslash U\|_\infty \qquad \forall f \in B_0$

holding for every $U \in \mathcal{V}$. It follows from lemma 7.6 that $B = C(V)$, that is,

(9) $\qquad A|V = C(V).$

This implies, as we have noted several times before in similar circumstances (via the Open Map Theorem), that the quotient norm in $A|V$ is equivalent to the uniform norm and so evidently A is bounded on the neighborhood V of x_0. Which we noted in the fourth paragraph of the proof, was all that had to be demonstrated.

<u>Corollary 7.9</u>

(Glicksberg [21]) Let X be a compact Hausdorff space, A a point-separating subalgebra of $C(X)$ which contains the constants and for which $A|F$ is uniformly closed in $C(F)$ for each closed $F \subset X$. Then $A = C(X)$.

<u>Proof</u>: We first show that A is normal. Let disjoint closed subsets F,K of X be given. If $x \in F$, $y \in K$, there exists $f \in A$ with $f(x) = 0$, $f(y) = 1$ ($1 \in A$, A separates points) and so there exist disjoint closed neighborhoods V_x of x and W_x of y such that

$$|f(V_x)| \le 1/4, \quad |1 - f(W_x)| \le 1/4.$$

By lemma 2.2 there exists a sequence of polynomials p_n which converges uniformly on $\{z \in \mathbb{C}: |z| \le 1/4\} \cup \{z \in \mathbb{C}: |1-z| \le 1/4\}$ to the characteristic function of $\{z \in \mathbb{C}: |z| \le 1/4\}$. Consequently $\{(p_n \circ f)|V_x \cup W_x\}$ is a sequence of elements of $A|V_x \cup W_x$ which converges uniformly on $V_x \cup W_x$ to the characteristic function of V_x. Since $A|V_x \cup W_x$ is closed in $C(V_x \cup W_x)$ by hypothesis, there exists $e \in A$ such that $e|V_x \cup W_x$ is the characteristic function of V_x.

Keeping $y \in K$ fixed, cover F with finitely many

$V_{x_1}, \ldots, V_{x_n}$ and let e_j be the element of A constructed above for the pair x_j, y so that $e_j(V_{x_j}) = 1$ and $e_j(W_j) = 0$ for some neighborhood W_j of y. Then the function $f_y = 1 - (1-e_1)(1-e_2)\ldots(1-e_n)$ belongs to A and is 1 on F and 0 on a neighborhood U_y of y. Cover K with $U_{y_1}, \ldots, U_{y_m}$ and take $f = f_{y_1} \ldots f_{y_m}$ to get a function in A which is 1 on F and 0 on K.

It is an immediate consequence of the Open Map Theorem and the fact (hypothesis) that $A|F$ is uniformly closed in $C(F)$ that the uniform norm in $A|F$, which is always dominated by the quotient norm, is in fact equivalent to it. It follows that the idempotents in A/kF are bounded in the quotient norm as the idempotents in $A|F$ are bounded (by 1) in the uniform norm. All the hypotheses of Theorem 7.8 are therefore satisfied by A in its uniform norm. (Note that, for example, taking $F = X$ in the hypothesis shows that A itself is uniformly closed in $C(X)$.) We conclude that $A = C(X)$.

Corollary 7.10

Let Y be a locally compact Hausdorff space, A a uniformly closed point-separating subalgebra of $C_0(Y)$ with the property that $A|F$ is uniformly closed in $C(F)$ for each compact $F \subset Y$. Then $A = C_0(Y)$.

Proof: Let X be a compact subset of Y. Then for any closed subset F of X we have $A|F$ closed in $C(F)$. Then $\mathbb{C} + A|F$ is also closed in $C(F)$. To see this let $c_n \in \mathbb{C}$, $f_n \in A|F$, $f \in C(F)$ and $c_n + f_n \to f$ uniformly on F. If $\{c_n\}$ is bounded then some subsequence $\{c_{n_j}\}$ converges to some $c \in \mathbb{C}$, so $\{f_{n_j}\}$ converges to $f - c$ and therefore $f - c \in A|F$. It follows that $f = c + (f - c) \in \mathbb{C} + A|F$. If on the other hand $\{c_n\}$ is not bounded, then some subsequence $\{|c_{n_j}|\}$ diverges to ∞ and $\frac{1}{c_{n_j}} f_{n_j} = \frac{1}{c_{n_j}}(c_{n_j} + f_{n_j}) - 1$ converges to -1 uniformly on F, so $-1 \in A|F$. It follows in this case that $\mathbb{C}+A|F = A|F$. The algebra $\mathbb{C} + A|X$ thus meets the hypotheses of Corollary 7.9 and so we conclude that $\mathbb{C} + A|X = C(X)$. But then $A|X$ is clearly a closed ideal in $C(X)$ and so by Corollary 1.8 is conjugate closed. Use the fact that A separates points and a finite covering argument to find $f_j \in A$ such that $g = |f_1|^2 + \ldots + |f_n|^2$ is never zero on X. As $A|X$ is conjugate closed, $h = g|X \in A|X$. Since $h > 0$, $1/h \in C(X)$ so $1 = \frac{1}{h} \cdot h \in A|X$, as the latter is an ideal. It follows $A|X = \mathbb{C} + A|X = C(X)$. Since X is an arbitrary compact subset of Y, the conclusion $A = C_0(Y)$ follows from Theorem 2.9.

Chapter VIII

CHARACTERIZATION OF $C(X)$ BY FUNCTIONS WHICH OPERATE

Our goal is a theorem of Katznelson (8.6) asserting that if the square root function on $\mathbb{R}^+$ operates (Definition 4.15) in a point-separating and conjugate-closed Banach subalgebra of $C(X)$, then that algebra is all of $C(X)$.

Lemma 8.1

Let A be a commutative Banach algebra with unit. If the set of idempotents in A is not bounded, then A contains an unbounded sequence of mutually orthogonal idempotents.

Proof: For each idempotent $h \in A$ set

(1) $N(h) = \sup\{\|hx\|_A : x \text{ idempotent}\} \in [0,\infty]$.

The hypothesis is that $N(1) = \infty$, 1 being the unit of A. For idempotents f,g,x

$$\|(f + g)x\|_A \leq \|fx\|_A + \|gx\|_A \leq N(f) + N(g).$$

If $f \cdot g = 0$, then $f + g$ is idempotent and taking the supremum on the left yields

$$N(f + g) \leq N(f) + N(g).$$

In particular

(2) f,g orthogonal idempotents & $N(f+g) = \infty \Rightarrow N(f)=\infty$ or $N(g)=\infty$.

Now choose idempotent g_1 with

$$\|g_1\|_A > 2.$$

Then let (recalling (2)) $h_1 = g_1$ or $h_1 = 1-g_1$ according as $N(1-g_1) = \infty$ or $N(g_1) = \infty$. It follows that

(3) $\qquad \|h_1\|_A > 1 \quad \& \quad N(1-h_1) = \infty.$

If mutually orthogonal idempotents $h_1,\ldots,h_{n-1}$ have been chosen with

(4) $\qquad \|h_k\|_A > k \quad (k = 1,2,\ldots,n-1) \quad \& \quad N(1 - \sum_{k=1}^{n-1} h_k) = \infty$

then pick idempotent x such that

(5) $\qquad \|(1 - \sum_{k=1}^{n-1} h_k)x\|_A > n + \|1 - \sum_{k=1}^{n-1} h_k\|_A.$

Then let (recalling (2)) $h_n = (1 - \sum_{k=1}^{n-1} h_k)x$ or $h_n = (1 - \sum_{k=1}^{n-1} h_k)(1-x)$ according as $N((1 - \sum_{k=1}^{n-1} h_k)(1-x)) = \infty$ or $N((1 - \sum_{k=1}^{n-1} h_k)x) = \infty$. It follows from (5) then that

(6) $\qquad \|h_n\|_A > n$

while

(7) $\qquad N(1 - \sum_{k=1}^{n} h_k) = \infty.$

Of course $h_j(1 - \sum_{k=1}^{n-1} h_k) = h_j - \sum_{k=1}^{n-1} h_j h_k = h_j - h_j = 0$ for $j = 1,2,\ldots,n-1$ so whichever choice is made for h_n we have

(8) $\qquad h_j \cdot h_n = 0 \qquad j = 1,2,\ldots,n-1.$

Now (6), (7) and (8) complete the inductive construction.

<u>Lemma 8.2</u>

Let X be a compact Hausdorff space, A a Banach algebra lying in $C(X)$, $\{h_n\}_{n=1}^{\infty}$ a sequence of mutually orthogonal idempotents of A, $\mu_n \geq 1$ $(n = 1,2,\ldots)$ real numbers.

Suppose that for every sequence c_n $(n = 1,2,\ldots)$ of non-negative numbers such that

$$\sum_{n=1}^{\infty} \mu_n c_n < \infty$$

A contains the function on X which takes the value c_n throughout the support of h_n for every n and is 0 elsewhere in X. Then there exists a finite constant K such that

$$\|h_n\|_A \le \mu_n K \qquad (n = 1,2,\ldots).$$

<u>Proof</u>: Let μ be the measure on $\mathbb{N}$ defined by $\mu(S) = \sum_{n\in S} \mu_n \in [0,\infty]$ for every $S \subset \mathbb{N}$. Let $\ell_1(\mu)$ denote the usual L_1-space on $\mathbb{N}$ with respect to this measure. As $\mu_n \ge 1$ it is clear that $\ell_1(\mu)$ is an algebra under pointwise multiplication and that the ℓ_1-norm is submultiplicative. For $f \in \ell_1(\mu)$ let $\Phi(f) = \sum_{n=1}^{\infty} f(n)h_n$, a function on X. Because of the mutual orthogonality of the h_n, the supports of the h_n are disjoint so there is no convergence question here. The hypothesis is that $\Phi(f) \in A$ for every $f \in \ell_1^+(\mu)$ and so by linearity for every $f \in \ell_1(\mu)$. Thus Φ is a homomorphism of $\ell_1(\mu)$ into A. An easy application of the Closed Graph Theorem shows that Φ is continuous. For if $f_n, f \in \ell_1(\mu)$ and $F \in A$ satisfy

(1) $\quad \|f_n - f\|_{\ell_1(\mu)} \to 0 \;\; \& \;\; \|\Phi(f_n) - F\|_A \to 0$

then since $\|\;\|_\infty \le \|\;\|_A$ (III, p. 1) we get

$$\|\Phi(f_n) - F\|_\infty \to 0$$

hence

(2) $\quad |\Phi(f_n)(x) - F(x)| \to 0$ for every $x \in X$.

Also from (1) follows

(3) $\quad |f_n(k) - f(k)| \to 0$ for each $k \in \mathbb{N}$.

Now each $\Phi(f_n)$ vanishes outside the union of the supports of the h_k and so by (2) F also has this property, while in the support of h_k each $\Phi(f_n)$ takes only the value $f_n(k)$ and so again from (2) F takes only the value $\lim_{n\to\infty} f_n(k) \overset{(3)}{=} f(k)$ there. It follows that $F = \Phi(f)$.

Now for each $k \in \mathbb{N}$ consider the characteristic function χ_k of $\{k\}$. We have evidently $\Phi(\chi_k) = h_k$ and $\|\chi_k\|_1 = \mu_k$. So

$$\|h_k\|_A = \|\Phi(\chi_k)\|_A \le \|\Phi\| \, \|\chi_k\|_1 = \mu_k \|\Phi\|.$$

So take for K the norm of the bounded linear operator Φ.

<u>Lemma 8.3</u>

Let X be an infinite compact Hausdorff space, A a Banach algebra lying in $C(X)$ which is normal and conjugate closed. If Φ on $[0,1)$ operates in A, then Φ is continuous.

<u>Proof</u>: As in the proof of lemma 4.22, because X is infinite there is a sequence of disjoint non-void open neighborhoods U_n in X. If $x_n \in U_n$ then x_n is not in the closure of the remaining x_k, so the normality of A plus Theorem 6.4(ii) provide $h_n \in A$ such that

(1) $\quad h_n(x_n) = 1,\ h_n(x_m) = 0 \quad m \neq n,\ h_n(x) \subset [0,1]$.

Now suppose that Φ is not continuous at some point

$a_0 \in [0,1)$, so that there exist $a_n \in [0,1)$ and $\delta > 0$ such that

(2) $\lim_{n\to\infty} a_n = a_0$

(3) $|\Phi(a_n) - \Phi(a_0)| \geq \delta \qquad n = 1,2,\ldots$.

Choose positive reals $b_1, b_2, \ldots$ so that

(4) $\sum_{k=1}^{\infty} b_k \|h_k\|_A < 1 - a_0$.

Then select $n_1 < n_2 < n_3 < \ldots$ so that

(5) $|a_{n_k} - a_0| < b_k$.

Define then (recalling that $1 \in A$ by normality)

$$h = a_0 + \sum_{k=1}^{\infty} (a_{n_k} - a_0)h_k.$$

The series converges in A-norm by (4) and (5). Also from (4), (5) and the fact (III, p.1) that $\| \ \|_\infty \leq \| \ \|_A$ and each h_n has range in $[0,1]$, we see that h has range in $[0,1)$. Since Φ operates we then have $\Phi\circ h \in A$. Let x_0 be a cluster point of the set $\{x_k\}_{k=1}^{\infty}$. Then since $\Phi\circ h$ is continuous

(6) $(\Phi\circ h)(x_0)$ is a cluster point of the numerical sequence $\{(\Phi\circ h)(x_k)\}_{k=1}^{\infty}$

and h being continuous

(7) $h(x_0)$ is a cluster point of the numerical sequence $\{h(x_k)\}_{k=1}^{\infty}$.

But $h(x_k) = a_{n_k}$ is immediate from (1) and the definition of h. And $a_{n_k} \to a_0$ so from (7)

(8) $h(x_0) = a_0$.

Thus (6) becomes the following contradiction to (3):

(9) $\Phi(a_0)$ is a cluster point of the numerical sequence $\{\Phi(a_{n_k})\}_{k=1}^{\infty}$.

Remark

For a related but deeper result see Corollary 9.2 of [17].

Lemma 8.4

Let X be a compact Hausdorff space, A a Banach algebra lying in $C(X)$ which is normal and conjugate closed. Let $F:[0,1) \to \mathbb{R}$ satisfy

$$(*) \qquad F(0) = 0 \quad \text{and} \quad \lim_{t \to 0^+} \frac{|F(t)|}{t} = \infty.$$

If F operates in A then the set of idempotents in A is bounded.

Proof: Suppose not. Let (lemma 8.1) h_n be an unbounded sequence of mutually orthogonal idempotents. Passing to a subsequence we may in view of (*) suppose without loss of generality that

$$(1) \qquad |F(t)| \geq nt \quad \text{for} \quad 0 \leq t \leq \|h_n\|_A^{-1} < n^{-1} \quad \text{for all} \quad n.$$

Let $\mu_n = n^{-1} \|h_n\|_A$ and let d_n be such that

$$(2) \qquad \sum_{n=1}^{\infty} d_n \mu_n < 1 \quad \text{and} \quad d_n \geq 0 \quad \text{for every} \quad n.$$

We have $0 \leq d_n/n \leq 1/(n\mu_n) = \|h_n\|_A^{-1}$, since by (2), $d_n \mu_n < 1$. So by (1)

$$F(\frac{d_n}{n}) \geq n\frac{d_n}{n} = d_n.$$

Now if X is finite our lemma is trivial. Otherwise by lemma 8.3 F is continuous and so there exist a_n such that

$$(3) \qquad 0 \leq a_n \leq \frac{d_n}{n} \quad \text{and} \quad F(a_n) = d_n$$

(by the Intermediate Value Theorem for continuous functions).

But

$$\sum_{n=1}^{\infty} a_n \|h_n\|_A \le \sum_{n=1}^{\infty} \frac{d_n}{n} \|h_n\|_A = \sum_{n=1}^{\infty} d_n \mu_n < 1$$

and consequently

$$f = \sum_{n=1}^{\infty} a_n h_n \in A$$

and $f(X) \subset [0,1)$. Therefore $F \circ f \in A$. Now

$$F \circ f(x) = F(f(x)) = F(\sum_{n=1}^{\infty} a_n h_n(x)) \qquad \forall x \in X$$

and so we see that the function $F \circ f$ of A takes the value $F(a_n) = d_n$ throughout the support of h_n and the value $F(0) = 0$ elsewhere in X. We conclude from lemma 8.2 that $\|h_n\|_A \mu_n^{-1}$ is bounded, contrary to the definition of μ_n.

<u>Corollary 8.5</u>

Let X be a compact Hausdorff space, A a Banach algebra lying in $C(X)$ which is normal and conjugate closed. Let $F: (-1,1) \to \mathbb{R}$ satisfy

$$F(0) = 0, \quad \lim_{t \to 0} \left|\frac{F(t)}{t}\right| = \infty.$$

If F operates in A, then $A = C(X)$.

<u>Proof</u>: If E is a closed subset of X and $A|E$ is given the quotient norm of A/kE, then $A|E$ is a Banach algebra lying in $C(E)$ and it is evidently normal and conjugate closed. We will show that F operates in $A|E$ and then conclude from the last lemma that in each such algebra the idempotents are bounded. It will follow from Theorem 7.8 that $A = C(X)$.

So let $f \in A|E$ and $f(E) \subset (-1,1)$. Pick $g \in A$ with $g|E = f$. Then $\frac{1}{2}(g + \overline{g}) \in A$ and $\frac{1}{2}(g + \overline{g})|E = f$ so we may assume g is real-valued. Then $g^{-1}(-1,1)$ is an open set which contains E. As A is normal there is a function φ in A which is 1 on E and 0 on the complement $X \backslash g^{-1}(-1,1) = |g|^{-1}[1,\infty)$ of this open set. According to Theorem 6.4 (ii) we may suppose that $\varphi(X) \subset [0,1]$. Consider $\varphi \cdot g \in A$. Evidently $\varphi \cdot g|E = f$. Moreover if $\varphi(x) \neq 0$ then $x \in g^{-1}(-1,1)$, so $|g(x)| < 1$ and $|\varphi(x)g(x)| \leq |g(x)| < 1$. Therefore $\varphi \cdot g(X) \subset (-1,1)$. As F operates in A we have that $F \circ (\varphi \cdot g) \in A$ and so

$$F \circ (\varphi \cdot g)|E \in A|E$$

that is, $F \circ (\varphi \cdot g|E) = F \circ f \in A|E$.

<u>Theorem 8.6</u>

(Katznelson [33]) Let X be a compact Hausdorff space and A a Banach algebra lying in $C(X)$ which is conjugate closed, contains the constants and separates the points of X. If the function $\sqrt{\ }$ operates in A then $A = C(X)$.

<u>Proof:</u> First notice that for any $f \in A$ the function $|f| = \sqrt{f\overline{f}}$ belongs to A. Therefore the function $F(t) = \sqrt{|t|}$ operates in A. This function obviously satisfies the conditions of the last corollary and so the desired result will follow from that corollary as soon as we show that A is normal.

If f is any real-valued function in A and $c \in \mathbb{R}$, then the function $f \vee c$ belongs to A since

$$f \vee c = \frac{|f-c|+f-c}{2} + c.$$

Next if x,y are distinct points of X, there is a real-valued $f \in A$ with $f(x) < 0$, $f(y) > 1$. Then the function $g = (f \wedge 1) \vee 0$ belongs to A and is 1 in the neighborhood $f^{-1}(1,\infty)$ of y and 0 in the neighborhood $f^{-1}(-\infty,0)$ of x. The argument in the second paragraph of the proof of Corollary 7.9 then establishes the normality of A.

Appendix

KATZNELSON'S IDEMPOTENT MODIFICATION TECHNIQUE

Prior to Badé and Curtis' proof of Corollary 6.16 in 1966, Katznelson [32] and [34] had proved a somewhat weaker version of it (the theorem below). Compare also Gorin [24]. Because Katznelson's technique in [34] involves a clever and intricate construction and because [34] was never published in a journal (and consequently has not perhaps reached as large an audience as it deserves) we are going to present the construction here.

From now on X shall be a compact Hausdorff space and A a Banach algebra lying in $C(X)$ which is boundedly ϵ-normal for some $\epsilon < \frac{1}{2}$ (hence for all $\epsilon > 0$ by lemma 6.2).

<u>Definition A.1</u>

Define a function $K_1: (0,\infty) \to [1,\infty)$ by $K_1(\epsilon) = 1 + \inf\{K$: If F_1, F_2 are disjoint closed subsets of X then A contains an ϵ-idempotent with respect to (F_1,F_2) of A-norm not greater than $K\}$.

By its definition then $K_1(\epsilon)$ satisfies

> For every $\epsilon > 0$ and disjoint closed subsets F_1,F_2 of X A contains an ϵ-idempotent with respect to (F_1,F_2) of A-norm less than $K_1(\epsilon)$.

The advantage of the function K_1 is that it is non-increasing. But in the next lemma, which is preparatory to the idempotent modification technique, it is advantageous to consider any

$$K: (0,\infty) \to [1,\infty)$$

such that K is non-increasing and $K(t) \geq K_1(t)$ for all $t > 0$.

Lemma A.2

Fix disjoint non-void closed subsets E_1, E_2 of X. Let $\epsilon_1 > 0$ and $h_1 \in A$ be an ϵ_1-idempotent with respect to (E_1, E_2) with

(0.1) $\max(\operatorname{Im} h_1) > 8\epsilon_1 > 0.$

Then there exists $h_2 \in A$ such that

(1.2) $\delta \overset{\text{def}}{=} \min\{\dfrac{\epsilon_1}{2K(\epsilon_1)}, \tfrac{1}{2}\}, \quad \epsilon_2 \overset{\text{def}}{=} \epsilon_1 + \dfrac{\delta \max(\operatorname{Im} h_1)}{4K(\delta)}$

(2.2) $\|h_2\|_A \leq \|h_1\|_A + \frac{1}{4}\max(\operatorname{Im} h_1)$

(3.2) $|h_2| \leq \epsilon_2$ on E_1

(4.2) $|h_2 - 1| \leq \epsilon_2$ on E_2

(5.2) $\max(\operatorname{Im} h_2) \leq [1 - \frac{1}{8K(\delta)}]\max(\operatorname{Im} h_1)$

(6.2) $\min(\operatorname{Im} h_2) \geq \min(\operatorname{Im} h_1) - \dfrac{\delta \max(\operatorname{Im} h_1)}{4K(\delta)}.$

Proof: Set $P_1 = \{x \in X: \operatorname{Im} h_1(x) \leq \frac{3}{8}\max(\operatorname{Im} h_1)\}$

$P_2 = \{x \in X: \operatorname{Im} h_1(x) \geq \tfrac{1}{2}\max(\operatorname{Im} h_1)\}$,

disjoint closed subsets of X. Let h_1^* be a δ-idempotent with respect to (P_1, P_2) with

(7) $\|h_1^*\|_A \leq K(\delta)$

(8) $|h_1^*| \leq \delta$ on P_1

(9) $|h_1^* - 1| \leq \delta$ on P_2.

Such h_1^* exist by definition of K_1 and the fact $K_1 \leq K$.

Set

(10) $$h_2 = h_1 - \frac{i \max(\mathrm{Im}\, h_1)}{4K(\delta)} h_1^* \in A.$$

Then (2.2) is clear from (7) [and (0.1)]. Notice that since h_1 is an ϵ_1-idempotent with respect to (E_1, E_2) we have on $E_1 \cup E_2$

$$|\mathrm{Im}\, h_1| = |\mathrm{Im}(h_1 - 1)| \le \epsilon_1 < \frac{1}{8} \max(\mathrm{Im}\, h_1) \quad \text{by (0.1).}$$

Therefore

(11) $$E_1 \cup E_2 \subset P_1.$$

Therefore (3.2) and (4.2) follow from the definition of ϵ_2 (1.2), the fact h_1 is an ϵ_1-idempotent and the fact $|h_1^*| \le \delta$ in $P_1 \supset E_1 \supset E_2$.

Now from (10)

(12) $$\mathrm{Im}\, h_2 = \mathrm{Im}\, h_1 - \frac{\max(\mathrm{Im}\, h_1)}{4K(\delta)} \mathrm{Re}\, h_1^*.$$

We analyze (5.2) and (6.2) in cases.

(I) In P_1:

$|\mathrm{Re}\, h_1^*| \le \delta$ by (8) and $\mathrm{Im}\, h_1 \le \frac{3}{8} \max(\mathrm{Im}\, h_1)$ by definition of P_1,

so from (12)

$$\begin{aligned}
\mathrm{Im}\, h_2 &\le \frac{3}{8} \max(\mathrm{Im}\, h_1) + \frac{\delta}{4K(\delta)} \max(\mathrm{Im}\, h_1) \\
&= \left[\frac{3}{8} + \frac{\delta}{4K(\delta)}\right] \max(\mathrm{Im}\, h_1) \\
&\le \left[\frac{3}{8} + \frac{1}{8K(\delta)}\right] \max(\mathrm{Im}\, h_1) \quad \text{since } \delta \le \tfrac{1}{2} \text{ by (1.2)} \\
&\le \left[\frac{3}{8} + \frac{1}{8}\right] \max(\mathrm{Im}\, h_1) \quad \text{since } K \text{ maps into } [1,\infty) \\
&= \tfrac{1}{2} \max(\mathrm{Im}\, h_1) \\
&\le \left[1 - \frac{1}{2K(\delta)}\right] \max(\mathrm{Im}\, h_1)
\end{aligned}$$

which gives (5.2) in this case.

On the other hand from (12) and $|\mathrm{Re}\ h_1^*| \le \delta$

$$\mathrm{Im}\ h_2 \ge \mathrm{Im}\ h_1 - \frac{\delta}{4K(\delta)}\max(\mathrm{Im}\ h_1)$$

$$(13)\qquad \mathrm{Im}\ h_2 \ge \min(\mathrm{Im}\ h_1) - \frac{\delta}{4K(\delta)}\max(\mathrm{Im}\ h_1).$$

(II) In P_2:

$|\mathrm{Re}\ h_1^*-1| \le \delta$ by (9) so $\mathrm{Re}\ h_1^* \ge 1-\delta$ and (12) gives

$$\mathrm{Im}\ h_2 \le \mathrm{Im}\ h_1 - \frac{(1-\delta)}{4K(\delta)}\max(\mathrm{Im}\ h_1)$$

$$\le [1 - \frac{1-\delta}{4K(\delta)}]\max(\mathrm{Im}\ h_1)$$

from which (5.2) follows in this case, upon recalling (1.2) that $\delta \le \frac{1}{2}$.

On the other hand in P_2 we have $\mathrm{Im}\ h_1 \ge \frac{1}{2}\max(\mathrm{Im}\ h_1)$ (by definition of P_2) and so (12) and $\mathrm{Re}\ h_1^* \le 1+\delta$ (9) give

$$(14)\qquad \mathrm{Im}\ h_2 \ge \tfrac{1}{2}\max(\mathrm{Im}\ h_1) - \frac{1+\delta}{4K(\delta)}\max(\mathrm{Im}\ h_1).$$

Since h_1 is an ϵ_1-idempotent with respect to (E_1,E_2) we have

$$\mathrm{Im}\ h_1 \le |h_1| \le \epsilon_1 \quad \text{on } E_1$$

and as E_1 is non-void by hypothesis this shows

$$\min(\mathrm{Im}\ h_1) \le \epsilon_1$$

$$\le \frac{1}{8}\max(\mathrm{Im}\ h_1) \quad \text{by (0.1)}$$

$$\le [\tfrac{1}{2} - \frac{1}{4K(\delta)}]\max(\mathrm{Im}\ h_1) \quad \text{since } K \ge 1$$

and so

$$\tfrac{1}{2}\max(\mathrm{Im}\ h_1) \ge \min(\mathrm{Im}\ h_1) + \frac{1}{4K(\delta)}\max(\mathrm{Im}\ h_1).$$

Putting this into (14) gives

$$(15)\qquad \mathrm{Im}\ h_2 \ge \min(\mathrm{Im}\ h_1) - \frac{\delta}{4K(\delta)}\max(\mathrm{Im}\ h_1).$$

(III) In $X\backslash P_1 \cup P_2$:

Here we have $\text{Im } h_1 < \frac{1}{2}\max(\text{Im } h_1)$ (definition of P_2) while $|\text{Re } h_1^*| \le \|h_1^*\|_\infty \le \|h_1^*\|_A \le K(\delta)$ by (7), so from (12)

$$\text{Im } h_2 \le \tfrac{1}{2}\max(\text{Im } h_1) + \frac{\max(\text{Im } h_1)}{4K(\delta)} \cdot K(\delta)$$

$$= [\tfrac{1}{2} + \tfrac{1}{4}]\max(\text{Im } h_1)$$

$$\le [1 - \frac{1}{4K(\delta)}]\max(\text{Im } h_1) \quad \text{since } K \ge 1,$$

giving (5.2) in this case.

On the other hand we also have $\text{Im } h_1 > \frac{3}{8}\max(\text{Im } h_1)$ here (definition of P_1) so (12) and $|\text{Re } h_1^*| \le K(\delta)$ give

$$\text{Im } h_2 \ge \frac{3}{8}\max(\text{Im } h_1) - \frac{\max(\text{Im } h_1)}{4K(\delta)} \cdot K(\delta)$$

$$= \frac{1}{8}\max(\text{Im } h_1) \ge \epsilon_1 \quad \text{by (0.1)}$$

$$\ge \min(\text{Im } h_1) \quad \text{as noted in the analysis of the previous case}$$

(16) $$\ge \min(\text{Im } h_1) - \frac{\delta}{4K(\delta)}\max(\text{Im } h_1).$$

Summarily (13), (15), (16) establish (6.2) and complete the proof of the lemma.

Now we iterate this construction: If

(0.2) $\max(\text{Im } h_2) > 8\epsilon_2$

then we apply the construction to (h_2, ϵ_2) in the role of (h_1, ϵ_1). Suppose for $k = 2,3,\ldots,n$ we have constructed $\epsilon_k > 0$ and functions $h_k \in A$ satisfying

(0.k-1) $\max(\text{Im } h_{k-1}) > 8\epsilon_{k-1} > 0$

(1.k) $$\epsilon_k = \epsilon_{k-1} + \frac{\delta \max(\text{Im } h_{k-1})}{4K(\delta)}$$

(2.k) $\qquad \|h_k\|_A \le \|h_{k-1}\|_A + \frac{1}{4}\max(\text{Im } h_{k-1})$

(3.k) $\qquad |h_k| \le \epsilon_k \quad \text{on} \quad E_1$

(4.k) $\qquad |h_k - 1| \le \epsilon_k \quad \text{on} \quad E_2$

(5.k) $\qquad \max(\text{Im } h_k) \le [1 - \frac{1}{8K(\delta)}]\max(\text{Im } h_{k-1})$

(6.k) $\qquad \min(\text{Im } h_k) \ge \min(\text{Im } h_{k-1}) - \frac{\delta}{4K(\delta)} \max(\text{Im } h_{k-1}).$

Induction on (5.k) $[k = 2,3,\ldots,n]$ gives

$$\max(\text{Im } h_k) \le [1 - \frac{1}{8K(\delta)}]^{k-1}\max(\text{Im } h_1)$$

(7.k) $$\le [1 - \frac{1}{8K(\delta)}]^{k-1} \|h_1\|_A \qquad k = 1,2,\ldots,n$$

using $\|\text{Im } h_1\|_\infty \le \|h_1\|_\infty \le \|h_1\|_A$ again (III, p. 1).

It follows by induction from (7.k) and (2.k) that

$$\|h_n\|_A \le \|h_1\|_A + \sum_{k=1}^{n-1} \frac{1}{4}[1 - \frac{1}{8K(\delta)}]^{k-1} \|h_1\|_A$$

$$\le \|h_1\|_A[1 + \sum_{k=1}^{\infty} \frac{1}{4}(1 - \frac{1}{8K(\delta)})^{k-1}]$$

$$= \|h_1\|_A[1 + 2K(\delta)]$$

(8.n) $$= \|h_1\|_A[1 + 2K(\min\{\tfrac{1}{2}, \frac{\epsilon_1}{2K(\epsilon_1)}\})].$$

It follows from (7.k) and (1.k) that

$$\epsilon_k \le \epsilon_{k-1} + \frac{\delta}{4K(\delta)}[1 - \frac{1}{8K(\delta)}]^{k-2} \|h_1\|_A.$$

By induction on this

$$\epsilon_n \le \epsilon_1 + \sum_{k=2}^{n} \frac{\delta}{4K(\delta)}[1 - \frac{1}{8K(\delta)}]^{k-2} \|h_1\|_A$$

$$\le \epsilon_1 + \frac{\delta \|h_1\|_A}{4K(\delta)} \sum_{k=2}^{\infty} [1 - \frac{1}{8K(\delta)}]^{k-2}$$

$$= \epsilon_1 + 2\delta \|h_1\|_A$$

whence

(1.n)' $\qquad \epsilon_n \le \epsilon_1[1 + \frac{\|h_1\|_A}{K(\epsilon_1)}] \qquad$ by (1.2).

From this and (3.n), (4.n) get

(9.n) $$|h_n| \le \epsilon_1[1 + \frac{\|h_1\|_A}{K(\epsilon_1)}] \quad \text{on } E_1$$

(10.n) $$|h_n - 1| \le \epsilon_1[1 + \frac{\|h_1\|_A}{K(\epsilon_1)}] \quad \text{on } E_2.$$

From (7.k) and (6.k) [also (0.k) is involved] we get

$$\min(\mathrm{Im}\, h_k) \ge \min(\mathrm{Im}\, h_{k-1}) - \frac{\delta}{4K(\delta)}[1 - \frac{1}{8K(\delta)}]^{k-2} \|h_1\|_A.$$

By induction on this

$$\min(\mathrm{Im}\, h_n) \ge \min(\mathrm{Im}\, h_1) - \sum_{k=1}^{n-1} \frac{\delta}{4K(\delta)}[1 - \frac{1}{8K(\delta)}]^{k-1} \|h_1\|_A$$

$$\ge \min(\mathrm{Im}\, h_1) - \frac{\delta\, \|h_1\|_A}{4K(\delta)} \sum_{k=1}^{\infty} [1 - \frac{1}{8K(\delta)}]^{k-1}$$

$$= \min(\mathrm{Im}\, h_1) - 2\delta\, \|h_1\|_A$$

(11.n) $$\ge \min(\mathrm{Im}\, h_1) - \frac{\epsilon_1 \|h_1\|_A}{K(\epsilon_1)} \quad \text{by (1.2).}$$

Now if in addition h_1 satisfies

(17) $$\|h_1\|_A \le K(\epsilon_1),$$

as will be the case in the applications to follow, then (9.n), (10.n) and (11.n) give

(12.n) $$|h_n| \le 2\epsilon_1 \quad \text{on } E_1$$

(13.n) $$|h_n - 1| \le 2\epsilon_1 \quad \text{on } E_2$$

(14.n) $$\min(\mathrm{Im}\, h_n) \ge \min(\mathrm{Im}\, h_1) - \epsilon_1.$$

Now while we do not claim that $\max(\mathrm{Im}\, h_k)$ is a decreasing function of k, it does follow from (7.k) and the fact (1.k) that the ϵ_k do not decrease with increasing k that (0.k-1) cannot hold for all positive integers k. Let us take for n the last k for which (0.k-1) holds. Then

(0.n) fails for

$$\epsilon_n = \epsilon_{n-1} + \frac{\delta \max(\mathrm{Im}\, h_{n-1})}{4K(\delta)},$$

that is,

$$\max(\mathrm{Im}\, h_n) \le 8\epsilon_n$$

$$\le 8\epsilon_1[1 + \frac{\|h_1\|_A}{K(\epsilon_1)}] \qquad \text{by (1.n)'}$$

$$\text{(15.n)} \qquad \le 16\epsilon_1 \qquad \text{by (17).}$$

Let us define for all $\epsilon > 0$

$$\text{(18)} \quad K_2(\epsilon) = K_1(\epsilon/16)[1 + 2K_1(\min\{\tfrac{1}{2}, \frac{\epsilon}{32K_1(\epsilon/16)}\})]$$

$$\text{(19)} \quad K_3(\epsilon) = 2K_2(\epsilon)[1 + 4K_2(\min\{2, \frac{\epsilon}{4K_2(\epsilon)}\})],$$

formidable functions to be sure, but they may be ignored for the present.

We now apply (twice) the idempotent modification technique just developed to prove

Lemma A.3

For any two disjoint non-void closed subsets F_1, F_2 of X and $\epsilon > 0$, A contains an ϵ-idempotent h with respect to (F_1, F_2) with $\|h\|_A \le K_3(\epsilon)$ and $\|\mathrm{Im}\, h\|_\infty \le 4\epsilon$.

Proof: Set $\epsilon_1 = \epsilon/16$ and pick an ϵ_1-idempotent h_1 with respect to (F_2, F_1) having $\|h_1\|_A \le K_1(\epsilon_1)$.

If $\max(\mathrm{Im}\, h_1) \le 8\epsilon_1$ set $h = h_1$ and have

$$\text{(20)} \quad \|h\|_A \le K_1(\epsilon/16) < K_2(\epsilon)$$

$$\text{(21)} \quad |h| \le \epsilon/16 \text{ on } F_2, \quad |h-1| \le \epsilon/16 \text{ on } F_1$$

$$\text{(22)} \quad \max(\mathrm{Im}\, h) \le \epsilon/2.$$

Otherwise we can apply the above construction with $E_1 = F_2$, $E_2 = F_1$ and $K = K_1$ to the pair (h_1, ϵ_1). Then set $h = h_n$ and get

(20)' $\|h\|_A \le K_2(\epsilon)$ by (8.n) and (18)

(21)' $|h| \le \epsilon/8$ on F_2, $|h-1| \le \epsilon/8$ on F_1 by (12.n),(13.n),(17).

(22)' $\max(\mathrm{Im}\ h) \le \epsilon$ by (15.n) and (17).

Now the modification technique truncates the imaginary part of the function from above. We would like to effect a truncation of the imaginary part of h from below so we consider applying this modification to $1-h$, since $\mathrm{Im}(1-h) = -\mathrm{Im}\ h$. However we must contend with the possibility that $1 \notin A$. But this is an inconsequential annoyance since A is boundedly ϵ-normal for every $\epsilon > 0$ and so A contains a function g such that

(23) $$\|g-1\|_\infty \le \epsilon/8$$

(that is, g is an $\epsilon/8$-idempotent for the pair $(\emptyset, X)$) and

(24) $$\|g\|_A \le K_1(\epsilon/8).$$

Set

(25) $$\tilde{h}_1 = g-h.$$

By (21) and (21)' we have

(26) $|\tilde{h}_1| \le |g-1| + |1-h| \le \epsilon/8 + \epsilon/8 = \epsilon/4$ on F_1

(27) $|\tilde{h}_1-1| \le |g-1| + |h| \le \epsilon/8 + \epsilon/8 = \epsilon/4$ on F_2

so that $\tilde{h}_1$ is an $\epsilon/4$-idempotent with respect to (F_1, F_2) which by (20), (20)' and (24) satisfies

$$\|h_1\|_A \le \|g\|_A + \|h\|_A \le K_1(\epsilon/8)+K_2(\epsilon) \le K_1(\epsilon/16)+K_2(\epsilon),$$

using the fact that K_1 is nonincreasing,

(28) $\|\tilde{h}_1\|_A \le 2K_2(\epsilon)$ by (18).

Note

$$\begin{aligned}\min(\operatorname{Im}\tilde{h}_1) &= \min(\operatorname{Im}(g - h)) \\ &\ge \min(\operatorname{Im}(-h)) - \max(\operatorname{Im} g) \\ &= \min(\operatorname{Im}(-h)) - \max(\operatorname{Im}(g - 1)) \\ &\ge \min(\operatorname{Im}(-h)) - \|g - 1\|_\infty \\ &= -\max(\operatorname{Im} h) - \|g - 1\|_\infty \\ &\ge -\epsilon - \|g - 1\|_\infty \quad \text{by (22) and (22)'} \end{aligned}$$

(29) $$\ge -\epsilon - \frac{\epsilon}{8} > -\frac{3\epsilon}{2} \quad \text{by (23).}$$

If

(30) $$\max(\operatorname{Im}\tilde{h}_1) \le 2\epsilon$$

then we set $\tilde{h} = \tilde{h}_1$ and have from (28)

(31) $$\|\tilde{h}\|_A \le 2K_2(\epsilon) \le K_3(\epsilon)$$

and from (26) and (27)

(32) $$|\tilde{h}| \le \epsilon/4 \text{ on } F_1, \ |\tilde{h}-1| \le \epsilon/4 \text{ on } F_2$$

while from (29)

(33) $$\min(\operatorname{Im}\tilde{h}) = \min(\operatorname{Im}\tilde{h}_1) \ge -\tfrac{3}{2}\epsilon$$

(34) $$\max(\operatorname{Im}\tilde{h}) = \max(\operatorname{Im}\tilde{h}_1) \le 2\epsilon \quad \text{by (30),}$$

and the lemma is proven. Otherwise

(35) $$\max(\operatorname{Im}\tilde{h}_1) > 2\epsilon$$

and we can apply the above construction with $E_1 = F_1$, $E_2 = F_2$, $K(\epsilon) = 2K_2(4\epsilon)$, $h_1 = \tilde{h}_1$ and $\epsilon_1 = \epsilon/4$. [Note that K_2 is nonincreasing (since K_1 is) and so this K is nonincreasing, while $K(\epsilon) \ge 2K_1(\epsilon/4) \ge K_1(\epsilon)$ as required for the construction.] Set $\tilde{h} = h_n$ and have

$$\|\tilde{h}\|_A \le \|\tilde{h}_1\|_A[1 + 2K(\min\{\tfrac{1}{2}, \frac{\epsilon}{8K(\epsilon/4)}\})] \quad \text{by (8.n)}$$

$$\le \|\tilde{h}_1\|_A[1 + 4K_2(\min\{2, \frac{\epsilon}{2K(\epsilon/4)}\})] \text{ by definition of K}$$

$$\le 2K_2(\epsilon)[1 + 4K_2(\min\{2, \frac{\epsilon}{4K_2(\epsilon)}\})] \text{ by (28) and definition of K}$$

(31)' $\qquad = K_3(\epsilon) \quad$ by (19).

On F_1 we have

$$|\tilde{h}| \le \epsilon/4[1 + \frac{\|h_1\|_A}{K(\epsilon/4)}] \qquad \text{by (9.n)}$$

$$\le \epsilon/4[1 + \frac{2K_2(\epsilon)}{K(\epsilon/4)}] \qquad \text{by (28)}$$

$$= \epsilon/2 \qquad \text{by definition of K here;}$$

and similarly with $\tilde{h}-1$ on F_2:

(32)' $\quad |\tilde{h}| \le \epsilon/2$ on F_1, $|\tilde{h}-1| \le \epsilon/2$ on F_2.

Next

$$\min(\mathrm{Im}\ \tilde{h}) \ge \min(\mathrm{Im}\ \tilde{h}_1) - \epsilon/4 \quad \text{by (14.n) and (17)}$$

$$\ge -\frac{3\epsilon}{2} - \frac{\epsilon}{4} > -2\epsilon \qquad \text{by (29)}$$

(33)' $\quad \min(\mathrm{Im}\ \tilde{h}) > -2\epsilon$.

Finally we have by definition of $\tilde{h}$ as h_n:

(34)' $\quad \max(\mathrm{Im}\ \tilde{h}) \le 16\frac{\epsilon}{4} = 4\epsilon \qquad$ by (15.n) and (17).

From (31) - (34) or (31)' - (34)' as the case may be, we see that $\tilde{h}$ does the required jobs.

Theorem A.4

(Katznelson [34] and Gorin [24]) If X is a compact Hausdorff space and A a Banach algebra lying in $C(X)$ which is boundedly ϵ-normal for some $\epsilon < \frac{1}{2}$, then $A = C(X)$.

Proof: It suffices to show $C^+_{\mathbb{R}}(X) \subset A$. For then $1 \in A$

and so $C_{\mathbb{R}}(X) = C^+_{\mathbb{R}}(X) + \mathbb{R} \subset A$ and then of course $C(X) = C_{\mathbb{R}}(X) + iC_{\mathbb{R}}(X) \subset A$.

Let c be any number satisfying

(1) $$0 < c < \frac{1}{4[1+K_3(1/8)]}$$

where K_3 is the function defined in (19) above. Consider then any $f \in C^+_{\mathbb{R}}(X)$. We may suppose $f \neq 0$. Define

(2) $$P_1 = \{x \in X: f(x) \geq 3/4 \|f\|_\infty\}$$

(3) $$P_2 = \{x \in X: f(x) \leq 1/2 \|f\|_\infty\}.$$

These are disjoint (as $\|f\|_\infty > 0$) closed subsets of X. Note that P_1 is non-void by definition of $\|f\|_\infty$ [and the facts $f \neq 0$, $f \geq 0$]. We consider the two possibilities for P_2.

(I) $P_2 \neq \emptyset$.

Then lemma A.3 is applicable and provides an $h \in A$ such that

(4) $$\|h\|_A \leq K_3(1/8)$$

(5) $$\|\text{Im } h\|_\infty \leq \tfrac{1}{2}$$

(6) $$|\text{Re } h| \leq 1/8 \quad \text{on } P_2$$

(7) $$|\text{Re } h-1| \leq 1/8 \quad \text{on } P_1.$$

Then define

(8) $$f_1 = c\|f\|_\infty h \in A.$$

We have from this and (4)

(9) $$\|f_1\|_\infty \leq \|f_1\|_A \leq cK_3(1/8)\|f\|_\infty$$

(10) $$< K_3(1/8)\|f\|_\infty.$$

On P_1 we have

$$f - \text{Re } f_1 \le \|f\|_\infty - c\|f\|_\infty \text{Re } h \overset{(7)}{\le} [1-\tfrac{7}{8}c]\|f\|_\infty < [1-\tfrac{5}{8}c]\|f\|_\infty$$

$$f - \text{Re } f_1 \overset{(2)}{\ge} \tfrac{3}{4}\|f\|_\infty - c\|f\|_\infty \text{Re } h \overset{(7)}{\ge} \tfrac{3}{4}\|f\|_\infty - c\|f\|_\infty \cdot \tfrac{9}{8} = [\tfrac{3}{4}-\tfrac{9}{8}c]\|f\|_\infty > 0$$

(11) $\quad |f - \text{Re } f_1| \le [1 - \tfrac{5}{8}c]\|f\|_\infty.$

On P_2 we have

$$f - \text{Re } f_1 \overset{(3)}{\le} \tfrac{1}{2}\|f\|_\infty - c\|f\|_\infty \text{Re } h \overset{(6)}{\le} [\tfrac{1}{2} + \tfrac{c}{8}]\|f\|_\infty \overset{(1)}{\le} [1-\tfrac{5}{8}c]\|f\|_\infty$$

$$f - \text{Re } f_1 \ge -\text{Re } f_1 = -c\|f\|_\infty \text{Re } h \overset{(6)}{\ge} -\tfrac{c}{8}\|f\|_\infty \overset{(1)}{\ge} -[1-\tfrac{5}{8}c]\|f\|_\infty$$

(12) $\quad |f - \text{Re } f_1| \le [1 - \tfrac{5}{8}c]\|f\|_\infty.$

On $X \backslash P_1 \cup P_2$

$$f - \text{Re } f_1 \overset{(2)}{\le} \tfrac{3}{4}\|f_1\|_\infty + \|f_1\|_\infty \overset{(9)}{\le} [\tfrac{3}{4} + cK_3(\tfrac{1}{8})]\|f\|_\infty \overset{(1)}{\le} [1-\tfrac{5}{8}c]\|f\|_\infty$$

$$f - \text{Re } f_1 \overset{(3)}{\ge} \tfrac{1}{2}\|f\|_\infty - \|f_1\|_\infty \overset{(9)}{\ge} [\tfrac{1}{2} - cK_3(1/8)]\|f\|_\infty \overset{(1)}{\ge} -[1-\tfrac{5}{8}c]\|f\|_\infty$$

(13) $\quad |f - \text{Re } f_1| \le [1 - \tfrac{5}{8}c]\|f\|_\infty.$

From (11), (12), (13)

(14) $\quad \|f - \text{Re } f_1\|_\infty \le [1 - \tfrac{5}{8}c]\|f\|_\infty.$

Then

$$\|f - f_1\|_\infty \le \|\text{Im } f_1\|_\infty + \|f - \text{Re } f_1\|_\infty$$

$$\|f - f_1\|_\infty \le \tfrac{c}{2}\|f\|_\infty + \|f - \text{Re } f_1\|_\infty \quad \text{by (5) and (8)}$$

$$\le \tfrac{c}{2}\|f\|_\infty + [1 - \tfrac{5}{8}c]\|f\|_\infty \quad \text{by (14)}$$

(15) $$= (1 - \tfrac{1}{8}c)\|f\|_\infty.$$

(II) $P_2 = \emptyset$.

Then since $f \ge 0$

(16) $\quad \tfrac{1}{2}\|f\|_\infty \le f \le \|f\|_\infty.$

Arguing as in lines 19 through 24 of the proof of lemma A.3, we see A contains an h with

(17) $\|h - 1\|_\infty \le \frac{1}{4}$ and

(18) $\|h\|_A \le K_1(\frac{1}{4})$.

Define then

(8)' $f_1 = \frac{4}{5}\|f\|_\infty h \in A$

and have from this and (18)

$$\|f_1\|_\infty \le \|f_1\|_A \le \tfrac{4}{5}K_1(1/4)\,\|f\|_\infty \le K_1(1/4)\,\|f\|_\infty \quad \le K_1(1/8)\|f\|_\infty$$

since K_1 is nonincreasing

(10)' $\|f_1\|_\infty \le K_3(1/8)\,\|f\|_\infty$ since $K_3 \ge K_2 \ge K_1$.

From (17) and (8)'

$$\frac{4}{5}\|f\|_\infty(1 - \frac{1}{4}) \le \text{Re } f_1 \le \frac{4}{5}\|f\|_\infty(1 + \frac{1}{4}).$$

It follows from this and (16) that

$$\begin{aligned} f - \text{Re } f_1 &\le \|f\|_\infty - \frac{4}{5}\|f\|_\infty(1 - \frac{1}{4}) \\ &= \frac{2}{5}\|f\|_\infty \\ f - \text{Re } f_1 &\ge \frac{1}{2}\|f\|_\infty - \frac{4}{5}\|f\|_\infty(1 + \frac{1}{4}) \\ &= -\frac{1}{2}\|f\|_\infty \end{aligned}$$

(19) $\|f - \text{Re } f_1\|_\infty \le \frac{1}{2}\|f\|_\infty$.

$$\begin{aligned} \|f - f_1\|_\infty &\le \|f - \text{Re } f_1\|_\infty + \|\text{Im } f_1\|_\infty \\ &= \|f - \text{Re } f_1\|_\infty + \frac{4}{5}\|f\|_\infty\,\|\text{Im } h\|_\infty \\ &\le \|f - \text{Re } f_1\|_\infty + \frac{1}{5}\|f\|_\infty && \text{by (17)} \\ &\le \frac{7}{10}\|f\|_\infty && \text{by (19)} \end{aligned}$$

(15)' $\qquad < (1 - \frac{1}{8}c)\,\|f\|_\infty.$

With (10), (10)' and (15), (15)' in hand, we could wrap-up the proof by appealing to Theorem 6.5. Alternatively we can apply this construction to $f - f_1$ in the role of f, etc. and produce

$$f_1, f_2, f_3, \ldots \in A$$

such that for every $n > 1$

$$(20) \quad \|f - (f_1 + \ldots + f_n\|_\infty \leq (1 - \tfrac{c}{8}) \|f - (f_1 + \ldots + f_{n-1})\|_\infty$$

$$(21) \quad \|f_n\|_A \leq K_3(1/8) \|f - (f_1 + \ldots + f_{n-1})\|_\infty .$$

Induction on (20) gives

$$(22) \quad \|f - (f_1 + \ldots + f_n)\|_\infty \leq (1 - \tfrac{c}{8})^n \|f\|_\infty$$

$$(23) \quad \|f_n\|_A \leq K_3(1/8)(1 - \tfrac{c}{8})^{n-1} \|f\|_\infty .$$

As $0 < 1 - \frac{c}{8} < 1$ it follows from (22) and (23) that

$$f = \sum_{n=1}^{\infty} f_n \in A.$$

REFERENCES

[1] E. Arenson, Certain properties of algebras of continuous functions, Soviet Math. 7 (1966), 1522-1524. MR 34, #6560.

[2] W. Badé, <u>The Banach Space C(S)</u>, Lecture Note Series No. 26, Aarhus University, 1971.

[3] W. Badé & P. Curtis Jr., Embedding theorems for commutative Banach algebras, Pac. J. Math. 18 (1966), 391-409. MR 34, #1878.

[4] A. Bernard, Une charactérisation de C(X) parmi les algèbres de Banach, C. R. Acad. Sci. Paris 267 (1968), 634-635. MR 38, #2601.

[5] A. Bernard, Algèbres ultraséparantes de fonctions continues, C. R. Acad. Sci. Paris 270 (1970), 818-819. MR 41, #826.

[6] A. Bernard, Fonctions qui opèrent sur Re A, C. R. Acad. Sci. Paris 271 (1970), 1120-1122. MR 42, #5048.

[7] A. Bernard, Espace des parties réelles des éléments d'une algèbre de Banach de fonctions, Jour. of Funct. Anal. (to appear).

[8] F. Birtel, ed. <u>Function Algebras</u>, Proceedings of an International Symposium held at Tulane University, 1965. Scott-Foresman (1966). MR 33, #4707.

[9] E. Bishop, A generalization of the Stone-Weierstrass Theorem, Pac. J. Math. 11 (1961), 777-783. MR 24,#3502.

[10] E. Bishop, A minimal boundary for function algebras, Pac. J. Math. 9 (1959), 629-642. MR 22, #191.

[11] L. de Branges, The Stone-Weierstrass Theorem, Proc. Amer. Math. Soc. 10 (1959), 822-824. MR 22, #3970.

[12] L. Brown, Subalgebras of L_∞ of the circle group, Proc. Amer. Math. Soc. 25 (1970), 585-587. MR 41, #4139.

[13] J. Cantwell, A topological approach to extreme points in function spaces, Proc. Amer. Math. Soc. 19 (1968), 821-825. MR 37, #4583.

[14] D. Chalice, Characterizations for approximately normal algebras, Proc. Amer. Math. Soc. 20 (1969), 415-419. MR 38, #2603.

[15] E. Čirka, Approximation of continuous functions by functions holomorphic on Jordan arcs in $\mathbb{C}^n$, Soviet Math. 7 (1966), 336-338. MR 34, #1563.

[16] B. Cole, One point parts and the peak point conjecture, thesis, Yale University (1968).

[17] P. Curtis, Jr., <u>Topics in Banach Spaces of Continuous Functions</u>, Lecture Note Series No. 25, Aarhus University, 1971.

[18] S. Fisher, The convex hull of the finite Blaschke products, Bull. Amer. Math. Soc. 74 (1968), 1128-1129. MR 38, #2316.

[19] I. Glicksberg, On two consequences of a theorem of Hoffman and Wermer, Math. Scand. 23 (1968), 188-192. MR 40, #3315.

[20] I. Glicksberg, Measures orthogonal to algebras and sets of antisymmetry, Trans. Amer. Math. Soc. 105 (1962), 415-435. MR 30, #4164.

[21] I. Glicksberg, Function algebras with closed restrictions, Proc. Amer. Math. Soc. 14 (1963), 158-161. MR 26, #616.

[22] I. Glicksberg, National Science Foundation Regional Conference on Recent Developments in Function Algebras, Univ. of Arkansas, Dept. of Mathematics, (1971).

[23] D. Goodner, The closed convex hull of certain extreme points, Proc. Amer. Math. Soc. 15 (1964), 256-258. MR 29, #455.

[24] E. Gorin, A property of the ring of all continuous functions on a bicompactum, Dokl. Akad. Nauk. SSSR 142 (1962), 781-784. MR 25, #2464.

[25] E. Gorin, Moduli of invertible elements in a normed algebra (Russian; English summary), Vestnik Moskov. Univ. Ser. I Mat. Meh. (1965), No. 5, 35-39. MR 32, #8206.

[26] L. Harris, Schwarz's lemma and the maximum principle in infinite dimensional spaces, Thesis, Cornell University (1969).

[27] L. Harris, Banach algebras with involution and Mobius transformations, Jour. of Funct. Anal. (to appear).

[28] K. Hoffman & A. Ramsey, Algebras of bounded sequences, Pac. J. Math. 15 (1965), 1239-1248. MR 33, #6442.

[29] K. Hoffman & J. Wermer, A characterization of C(X), Pac. J. Math. 12 (1962), 941-944. MR 27, #325.

[30] H. Ishikawa, J. Tomiyama, J. Wada, On the local behavior of function algebras, Tôhoku Math. J. 22 (1970), 48-55. MR 41, #4249.

[31] K. deLeeuw & Y. Katznelson, Functions that operate on non-self-adjoint algebras, Jour. d'Anal. Math. 11 (1963), 207-219. MR 28, #1508.

[32] Y. Katznelson, A characterization of all continuous functions on a compact Hausdorff space, Bull. Amer. Math. Soc. 66 (1960), 313-315. MR 22, #12404.

[33] Y. Katznelson, Sur les algèbres dont les éléments non-négatifs admettent des racines carrées, Ann. Scient. Éc. Norm. Sup. 77 (1960), 167-174. MR 22, #12403.

[34] Y. Katznelson, Characterization of C($\mathfrak{M}$), Technical Note No. 25, Air Force Office of Scientific Research, Hebrew University, Jerusalem (1962).

[35] R. Kaufman, Operators with closed range, Proc. Amer. Math. Soc. 17 (1966), 767-768. MR 34, #4908.

[36] R. McKissick, Existence of a nontrivial normal function algebra, Thesis, Massachusetts Institute of Technology (1963).

[37] R. McKissick, A nontrivial normal sup norm algebra, Bull. Amer. Math. Soc. 69 (1963), 391-395. MR 26, #4166.

[38] R. Mullins, The algebra of continuous functions characterized by a countable number of restrictions, Archiv der Math. 21 (1970), 66-68. MR 41, #8996.

[39] S. Negrepontis, On a theorem of Hoffman and Ramsey, Pac. J. Math. 20 (1967), 281-282. MR 35, #741.

[40] R. Phelps, Extreme points in function algebras, Duke J. Math. 32 (1965), 267-278. MR 30, #3890.

[41] W. Rudin, Real and Complex Analysis, McGraw-Hill, New York (1966). MR 35, #1420.

[42] W. Rudin, Convex combinations of unimodular functions, Bull. Amer. Math. Soc. 75 (1969), 795-797. MR 39, #3239.

[43] W. Rudin, Continuous functions on compact spaces without perfect subsets, Proc. Amer. Math. Soc. 8 (1957), 39-42. MR 19, p. 46.

[44] B. Russo & H. Dye, A note on unitary operators in C^*-algebras, Duke J. Math. 33 (1966), 413-416. MR 33, #1750.

[45] Z. Semadeni, Banach Spaces of Continuous Functions, Polish Scientific Publishers, Warsaw (1971).

[46] S. Sidney, Notes on Rational Approximation, Yale University, Dept. of Mathematics, (1970).

[47] S. Sidney & L. Stout, A note on interpolation, Proc. Amer. Math. Soc. 19 (1968), 380-382. MR 36, #6944.

[48] R. Sine, On a paper of Phelps', Proc. Amer. Math. Soc. 18 (1967), 484-486. MR 35, #2152.

[49] L. Stout, The Theory of Uniform Algebras, Bogden & Quigley, Tarrytown-on-Hudson (1971).

[50] J. Wermer, The space of real parts of a function algebra, Pac. J. Math. 13 (1963), 1423-1426. MR 27, #6152.

[51] I. Wik, On linear dependence in closed sets, Arkiv för Math. 4 (1961), 209-218. MR 23, #A2707.

[52] D. Wilken & T. Gamelin, Closed partitions of maximal ideal spaces, Ill. J. Math. 13 (1969), 789-795. MR 40, #4767.

SYMBOL INDEX

supp μ, 6

$|| \; ||_\infty$, $|| \; ||_X$, $|| \; ||_A$, 1

$\mathbb{N}$ = natural numbers, 1

χ_B = characteristic function of set B

Re f = real part of f

Re A = {Re f : f $\in$ A}

$\mathbb{R}$, $\mathbb{R}^+$, 1

$\sqrt{\ }$ = non-negative square root defined on $[0,\infty)$

T* = adjoint of the bounded operator T

Y* = space of continuous linear functionals on the Banach space Y

Z, Z^+, 1

$\perp$, 6

$\forall$ = universal quantifier

$\exists$ = existential quantifier

$\setminus$ = set theoretic difference

$\circ$ = functional composition

$*$ = convolution : $\varphi * \psi(x) = \int_{\mathbb{R}} \varphi(x-y)\psi(y)dy$

SUBJECT INDEX